IMMERS

AERIALISTS

LONG-LEGGED WADERS

SMALLER WADERS

FOWL-LIKE BIRDS

BIRDS OF PREY

NONPASSERINE LAND BIRDS

PASSERINE (PERCHING) BIRDS

THE PETERSON FIELD GUIDE SERIES®

LARGE FORMAT EDITION

A FIELD GUIDE TO THE

BIRDS

EASTERN AND CENTRAL
NORTH AMERICA

Text and Illustrations by

ROGER TORY PETERSON

Maps by

VIRGINIA MARIE PETERSON

SPONSORED BY
THE NATIONAL WILDLIFE FEDERATION AND
THE ROGER TORY PETERSON INSTITUTE

HOUGHTON MIFFLIN COMPANY
BOSTON   NEW YORK

CONSERVATION NOTE

Birds undeniably contribute to our pleasure and standard of living. But they also are sensitive indicators of the environment, a sort of "ecological litmus paper," and hence more meaningful than just chickadees and cardinals to brighten the suburban garden, grouse and ducks to fill the sportsman's bag, or rare warblers and shorebirds to be ticked off on the birder's checklist. The observation of birds leads inevitably to environmental awareness.

Help support the cause of wildlife conservation by taking an active part in the work of the National Audubon Society (700 Broadway, New York, NY 10003), the National Wildlife Federation (11100 Wildlife Center Drive, Reston, VA 20190), The Nature Conservancy (4245 North Fairfax Dr., Suite 100, Arlington, VA 22203-1606), and your local Audubon or Natural History Society. On the international level, don't forget the World Wildlife Fund (1250 24th Street N.W., Washington, D.C. 20037). These and other conservation organizations merit your support.

Text copyright © 1999 by The Marital Trust B u/w
Roger Tory Peterson and Virginia Peterson
Illustrations copyright © 1980 by Roger Tory Peterson
Maps copyright © 1999 by Virginia Peterson

All rights reserved

For information about permission to reproduce selections from
this book, write to Permissions, Houghton Mifflin Company,
215 Park Avenue South, New York, New York 10003.

Visit our Web site: www.hmhco.com.

PETERSON FIELD GUIDES and PETERSON FIELD GUIDE SERIES
are registered trademarks of Houghton Mifflin Company.

Library of Congress Cataloging in Publication Data
Peterson, Roger Tory, 1908–1996
A field guide to the birds : eastern and central North America /
text and illustrations by Roger Tory Peterson ; maps by Virginia
Marie Peterson. —Large format ed.
p.   cm.—(The Peterson field guide series)
"Sponsored by the National Wildlife Federation and the Roger Tory
Peterson Institute."
ISBN 0-395-96371-0
ISBN 978-0-395-96371-5

1. Birds—North America. I. Title. II. Series.
QL681.P446   1999
598'.097—dc21   99-20466
CIP

Consultant on text: Dr. Noble Proctor
Consultant on maps: Paul Lehman
Consultant on overview: Pete Dunne
Based upon *A Field Guide to the Birds*,
© 1980 by Roger Tory Peterson.
Fourth Edition,

Printed in China
SCP   11 10 9 8

to the memory of

CLARENCE E. ALLEN

and

WILLIAM VOGT

CONTENTS

INTRODUCTION

In 1934 my husband, Roger Tory Peterson, published his first Field Guide, which covered the birds east of the 90th meridian in North America. This book was designed so that live birds could be readily identified at a distance by their "field marks" without resorting to the bird-in-hand characteristics that the early collectors relied on. During the last half-century the binocular and the spotting scope have replaced the shotgun.

The "Peterson System," as it now is called, is based on patternistic drawings with arrows that pinpoint the key field marks. These rather formal schematic illustrations and the direct comparisons between similar species are the core of the system, a practical method that has gained universal acceptance not only on this continent but also in Europe where Peterson Field Guides now exist in 12 languages. This system, which is, in a sense, a pictorial key based on readily noticed visual impressions rather than on technical features, has been extended to other branches of natural history and there are now more than four dozen titles in the Peterson Field Guide Series.

After the death of my husband on July 28, 1996, I was approached by Houghton Mifflin Company to consider the idea of a new format for the *Field Guide to Eastern Birds*. The change would include enlarging the size of the plates, shortening the text, and placing updated range maps close to the birds.

I felt the idea of the new format had merit, and I proceeded to find the most knowledgeable consultants. Those chosen were friends close to Roger, and they shared his lifetime commitment to the knowledge and protection of birds. I owe a great debt of gratitude to Dr. Noble Proctor for his expert assistance on the text, to Paul Lehman for his broad knowledge in updating the range maps, and to Pete Dunne for his final overview. Roger, I know, would be honored and pleased with their contributions.

There are 395 maps in this book. Dr. Peterson and I researched all of the original map material. My previous research background involved work for the U.S. Coast Guard Research and Development Center, where I developed critical methods for identifying oil spills by means of infrared spectroscopy. This led to my writing the original *Infrared Field Manual for Oil Spill Identification*.

This new large format edition is another step in the progression of this

Field Guide. Years ago Dr. Peterson concluded that for comparative purposes the ideal number of species per color plate would be about 4 (rather than 10 to 12 as in the previous editions), but the cost factor prohibited this ideal format when the first Field Guide was published. In fact, there were only 4 plates in color and 26 in black and white in the 1934 edition. In 1939, 4 more black-and-white plates were added. When the book was extensively revised in 1947 and its scope extended to the 100th meridian, all the old plates were retired and replaced by 60 new ones, 36 in color. The success of the Field Guide with its well-tested practical system has grown steadily over the years and the economics of distribution as well as technical advances in printing now make it possible to surmount earlier restraints.

The Field Guide user will find a major format change that will be particularly helpful: all maps are now with the species descriptions. Over the years many birders had urged Dr. Peterson to arrange text, plates, and maps in this way, and now we are pleased to be able to provide this means of quicker reference in this large format edition.

AREA OF THIS FIELD GUIDE: Roughly, this guide covers North America east of the 100th meridian as shown in the map on the inside of the front cover. Rather than a restrictive political boundary, an ecological one is more practical. In the U.S. the logical division of the avifauna is in the belt between the 100th meridian (midway across Oklahoma, Kansas, Nebraska, and the Dakotas) and the edge of the Rockies. This is by no means a sharp division, but people living in that ecological area of overlap will find that *A Field Guide to Western Birds* covers all species they are likely to encounter. In a general way, eastern birds follow the valleys west while western forms edge eastward along the more arid uplands. In Canada eastern influences extend much farther west, bridging the gap to the Rockies via the conifer forests north of the plains.

The birds resident in the eastern third of Texas are adequately covered in this Field Guide. Not so those in the Rio Grande Valley and along the lower and central coasts of Texas, where many western species reach their eastern outposts and a few Mexican species occur. These will be found in *A Field Guide to the Birds of Texas.* Texas is the only state with its own Field Guide.

THE MAPS AND RANGES OF BIRDS: A number of species have been added to the avifauna of eastern North America since the previous edition of the Field Guide was published in 1980. Taxonomic splits have resulted in such "new" species as Saltmarsh and Nelson's Sharp-tailed Sparrows and Eastern and Spotted Towhees. Successful introductions of some species have resulted in self-sustaining, growing populations of Trumpeter Swans, Canary-winged (Yellow-chevroned) Parakeets, and Eurasian Collared-Doves (the latter was introduced to the Bahamas, then arrived in the U.S. on its own). In addition, a good number of new vagrant species—out-of-range visitors from far-away lands—continue to be found. Some species that were formerly thought to occur only exceptionally have, over the past several decades, become much more regular visitors and sometimes even local breeders. It is not always certain if such increases are the result of the birds' populations increasing and/or spreading, or if they merely reflect better observer coverage and advances in field identification skills.

The ranges of many species have changed markedly over the past 50 or more years. Some are expanding because of protection given them, changing habitats, bird feeding, or other factors. Some "increases" may simply be the result of more Field Guide–educated birders being in the field, helping to more thoroughly document bird populations and distributions. Other avian species have diminished alarmingly and may have been extirpated from major parts of their range. The primary culprit here has been habitat loss, although other factors such as increased competition or predation from other species may sometimes be involved. Species that are in serious decline in eastern North America run the gamut, from the American Bittern to the Loggerhead Shrike and Bewick's Wren.

KEY TO RANGE MAPS

Red: summer range

Blue: winter range

Purple: year-round range

Red dash line: approximate limits of irregular summer range and/or post-breeding dispersal

Blue dash line: approximate limits of irregular winter range

Purple dash line: approximate limits of irregular year-round range

DRAWINGS VS. PHOTOGRAPHS: Because of the increasing sophistication of birders Dr. Peterson leaned more toward detailed portraiture in the illustrations while trying not to lose the patternistic effect developed in previous editions. A drawing can do much more than a photograph to emphasize the field marks. A photograph is a record of a fleeting instant; a drawing is a composite of the artist's experience. The artist can edit out, show field marks to best advantage, and delete unnecessary clutter. He can choose position and stress basic color and pattern unmodified by transitory light and shade. A photograph is subject to the vagaries of color temperature, make of film, time of day, angle of view, skill of the photographer, and just plain luck. The artist has more options and far more control even though he may at times use photographs for reference. This is not a diatribe against photography; Dr. Peterson was an obsessive photographer as well as an artist and fully aware of the differences. Whereas a photograph can have a living immediacy, a good drawing is really more instructive.

SUBSPECIES: These simply represent subdivisions within the geographic range of a species. They are races, usually determined by morphological characteristics such as slight differences in measurements, shades of color, etc. These subtle subdivisions in many cases can be distinguished only by examination of a museum series of skins, or, in recent years, by DNA recombination studies of populations that are isolated from one another. Subspecies are important to students of bird distribution and evolution. They have also proven to be important for wildlife management, as seen with the Cape Sable and Merritt Island subspecies of the Seaside Sparrow. For many birders it is an enjoyable challenge to identify some of these subspecies afield, and records of occurrence beyond the normal range could prove to be of value in the future as populations continue to change over time. An example of the value of such studies can be seen in a group such as the flycatchers, in which several subspecies of the past have now proven to be full species. Birders who recorded the races involved have provided valuable information on occurrence and range boundaries for the newly separated species. Other well-known examples are the "Myrtle" and "Audubon's" subspecies of the Yellow-rumped Warbler. The Dark-eyed Junco has three easily recognized races, the "Slate-colored," "Oregon," and "White-winged" Juncos. Studies of these often subtle differences will help enhance the overall field identification skills of the birder. A copy of the A.O.U. (American Ornithologists' Union) Check-list is an excellent resource that gives a detailed breakdown of races and ranges. Earlier editions of this Field Guide also detailed subspecies, but those pages have been put to other uses in this guide.

BIRD SONGS AND CALLS: Not everything useful for identifying birds can be crammed into a pocket-sized Field Guide. In the species accounts Dr. Peterson included a brief entry on voice and tried to give the reader some handle on the songs or calls he or she hears. Authors of bird books have attempted with varying success to fit songs into syllables, words, and phrases. Musical notations, comparative descriptions, and even ingenious systems of symbols have also been employed. But since the advent of sound recording these older techniques have been eclipsed. A visual spinoff of the tape recording is the sonogram, but most people are not technologically oriented enough to interpret it easily.

Use the *Field Guide to Bird Songs,* which is arranged to match this edition of the Field Guide to Birds. It comprises two CDs. This comprehensive collection of sound recordings includes the calls and songs of more than 200 land and water birds—a large percentage of all the species found in eastern and central North America. They were recorded and prepared under the direction of Dr. James Gulledge of the Laboratory of Ornithology, Cornell University. In addition, *Birding By Ear: Eastern and Central North America* and *More Birding By Ear,* by Richard K. Walton and Robert W. Lawson are available in both CDs and cassettes. To prepare yourself for your field trips play the CDs, then read the descriptions in the Field Guide for useful clues. In learning bird voices (and some birders do 90 percent of their field work by ear) there is no substitute for the actual sounds.

ACKNOWLEDGMENTS: It was William Vogt, the first editor of *Audubon* magazine, who suggested that my husband put together a Field Guide using his methods of teaching field identification. He was the spark plug. Dr. Peterson had already written articles on the subject for *Nature* magazine (gulls) and *Field and Stream* (ducks). Vogt and Dr. Peterson enjoyed many field trips together during Dr. Peterson's art school days and when they were on the staff of the National Audubon Society. Vogt was to distinguish himself later as one of the early gurus of the environmental movement.

There was no efficient bird guide in the modern sense in Dr. Peterson's youth. He used the little checkbook-sized *Reed's Bird Guide,* but it was not until he met the young members of the Bronx County Bird Club—Allan Cruickshank, Richard Herbert, Joseph Hickey, Irving Kassoy, and the Kuerzi brothers—that Dr. Peterson really learned his birds. The man to whom they all owed their inspiration and expertise was Ludlow Griscom of the Museum of Comparative Zoology at Cambridge, Massachusetts. He was the court of last resort in matters of field identification and to him Dr. Peterson always turned for a final appraisal of difficult species and knotty problems. Throughout the preparation of the early editions of the Field Guide Griscom unselfishly gave Dr. Peterson the benefit of his long and varied field experience. Charles A. Urner of Elizabeth, New Jersey, whose specialty was the water birds, also graciously criticized the original manuscript. So did Francis H. Allen of Boston, who was Dr. Peterson's first editor. He contributed many valuable notes and was responsible for a complete perusal and polishing of the original text as well as that of the first and second revisions.

In the original Field Guide my husband lists more than 100 correspondents that had input into the original version. Their efforts remain in the modified text. While maintaining and updating ranges for the maps a consistent flow of information was sent from each of the states covered by this text. I would like to thank the following for their kindness, time, and effort. Many have been involved with their state's or province's book, atlas project, or published bird atlas. Their knowledgeable input makes the maps as up to date as possible for this large format edition.

ALABAMA: T. Imhof, G. Jackson; **ARKANSAS:** D. Catanzaro, M. Parker, W. Shepherd, K. Smith, M. White; **CONNECTICUT:** T. Baptist, L. Bevier, H. Golet, N. Proctor, J. Zeranski; **DELAWARE:** M. Barnhill, L. Fleming, G. Hess, R. West; **FLORIDA:** B. Anderson, W. Robertson, H. Stevenson, G. Woolfendon; **GEORGIA:** A. Ashley, W. Baker, G. Beaton, B. Bergstrom, K. Blackshaw, B. Blakeslee, M. Chapman, L. Davenport Jr., H. DiGioia, J. Greenberg, D. Guynn Jr., M. Harris, J. Hitt, M. Hodges, M. Hopkins Jr., W. Hunter, M. Oberle, J. Ozier, J. Paget, J. Parrish, S. Pate, T. Patterson, T. Schneider, P. Sykes, S. Willis, B. Winn; **ILLINOIS:** D. Bohlen, R. Chapel, V. Kleen, W. Serafin, E. Walters, W. Zimmerman; **INDIANA:** K. Brock, J. Castrale et al, E. Hopkins, C. Keller; **IOWA:** J. Dinsmore, J. Giglerano, B. Hoyer, L. Jackson, K. Kane, T. Kent, P. Lohmann, C. Thompson; **KANSAS:** C. Ely, C. Hobbs, M. Robbins, S. Seltman, M. Thompson, J. Zimmerman;

KENTUCKY: B. Palmer-Ball Jr.; **LOUISIANA:** J. V. Remsen, M. Swan, D. Wiedenfeld; **MAINE:** P. Adamus, E. Pierson et al; **MARYLAND:** C. Robbins; **MASSACHUSETTS:** P. Alden, W. Petersen, R. Veit; **MICHIGAN:** R. Adams Jr., C. Black, R. Brewer, G. McPeek, C. Smith; **MINNESOTA:** K. Eckert, R. Janssen; **MISSISSIPPI:** J. Jackson, T. Scheifer, J. Wilson; **MISSOURI:** D. Easterla, C. Hobbs, B. Jacobs, P. McKenzie, M. Robbins, J. Wilson; **MONTANA:** T. McEneaney; **NEBRASKA:** P. Johnsguard, W. Mollhoff, L. Paddleford, R. Silcock; **NEW JERSEY:** P. Dunne, V. Elia, D. Hughs, P. Lehman, J. Walsh; **NEW HAMPSHIRE:** C. Foss; **NEW MEXICO:** J. Oldenettel; **NEW YORK:** R. Andrle, P. Buckley, J. Carroll, E. Levine; **NORTH CAROLINA:** C. Brimley, H. Brimley, H. Davis, J. Fussell, D. Lee, T. Pearson, N. Siebenheller, W. Siebenheller, D. Wray; **NORTH DAKOTA:** G. Berkley, R. Martin, R. Stewart; **OHIO:** B. Peterjohn, D. Rice, L. Rosche, W. Zimmerman; **OKLAHOMA:** W. Carter, J. Grzybowski, G. Schnell, G. Sutton, D. Wood; **PENNSYLVANIA:** D. Brauning; **RHODE ISLAND:** R. Enser; **SOUTH CAROLINA:** J. Cely, M. Dodd, S. Gauthreaux, T. Murphy, W. Post, P. Wilkinson; **SOUTH DAKOTA**: C. Lippincott, R. Peterson, The South Dakota Ornithologist's Union; **TENNESSEE:** B. Linsey, C. Nicholson, J. Robinson, J. Wilson; **TEXAS:** K. Arnold, R. Baker, K. Benson, G. Blackloack, E. Kincaid, G. Lasley, M. Lockwood, H. Oberholser, J. Rappole; **VERMONT:** D. Kibbe, S. Laughlin; **VIRGINIA:** T. Dalmas, Virginia Society of Ornithology; **WEST VIRGINIA:** A. R. Buckelew Jr., G. A. Hall; **WISCONSIN:** D. Flaspohler, S. Matteson, S. Robbins Jr.; **BRITISH COLUMBIA:** R. Cambell et al; **LABRADOR AND NEWFOUNDLAND:** C. Brown, K. Knowles, P. Linegar, B. Mactavish, W. Montevecchi, J. Pratt, P. Ryan, J. Selno, J. Wells; **MANITOBA:** B. Carey, R. Coes, C. Curtis, G. Holland, B. Knudsen, W. Neeley, P. Taylor; **NORTHWEST TERRITORY:** W. E. Godfrey; **NEW BRUNSWICK:** B. Dalzell; **NOVA SCOTIA:** B. Dalzell, A. Erskine, I. McLaren, R. Tufts; **ONTARIO:** M. Cadman, P. Eagles, F. Helleiner; **QUEBEC:** Y. Aubry, A. Cyr, N. David, J. Gauthier, J. Larivee; **ST. PIERRE AND MIQUELON:** R. Etcheberry; **SASKATCHEWAN:** A. R. Smith; **MEXICO:** S. N. G. Howell, S. Webb.

ADDITIONAL HELP: P. Bacinski, W. Boyle, W. Burt, V. Emanuel, K. Garrett, F. Gill, K. Kaufman, K. Parkes, A. Poole, J. V. Remsen, J. Rowlett, P. Stellenheim, W. Sladen, R. Sundell, P. Vickery.

Putting a Field Guide together so that everything fits is a challenge comparable to a jigsaw puzzle or a game of chess. A myriad of details from writing letters, finding references, and coordinating meetings were put into the capable hands of my secretary, Elaine Lillis, and my studio assistant, Elizabeth Gentile. Their professionalism and support has been invaluable to make this book come to fruition. And to Barry Estabrook, Lisa White, Anne Chalmers, and Evelyn Defossez at Houghton Mifflin who provided technical support and guidance to see this project through I extend my great appreciation.

— *Virginia Marie Peterson*

HOW TO IDENTIFY BIRDS

Veteran birders will know how to use this book. Beginners, however, should spend some time becoming familiar in a general way with the illustrations. They are not arranged in systematic or phylogenetic order as in most ornithological works but are grouped in 8 main visual categories:

(1) **Swimmers**—Ducks and ducklike birds
(2) **Aerialists**—Gulls and gull-like birds
(3) **Long-legged Waders**—Herons, cranes, etc.
(4) **Smaller Waders**—Plovers, sandpipers, etc.
(5) **Fowl-like Birds**—Grouse, quail, etc.
(6) **Birds of Prey**—Hawks, eagles, owls
(7) **Nonpasserine Land Birds**
(8) **Passerine (Perching) Birds**

Within these groupings it will be seen that ducks do not resemble loons; gulls are readily distinguishable from terns. The needlelike bills of warblers immediately differentiate them from the seed-cracking bills of sparrows. Birds that could be confused are grouped together when possible and are arranged in identical profile for direct comparison. The arrows point to outstanding "field marks" which are explained opposite.

WHAT IS THE BIRD'S SIZE?

Acquire the habit of comparing a new bird with some familiar "yardstick"—a House Sparrow, a robin, a pigeon, etc., so that you can say to yourself, "smaller than a robin; a little larger than a House Sparrow." The measurements in this book represent lengths in inches from bill tip to tail tip of specimens on their backs as in museum trays. However, specimen measurements vary widely depending on the preparator, who may have stretched the neck a bit.

WHAT IS ITS SHAPE?
Is it plump like a Starling (left) or slender like a cuckoo (right)?

WHAT SHAPE ARE ITS WINGS?
Are they rounded like a Bobwhite's (left) or sharply pointed like a Barn Swallow's (right)?

WHAT SHAPE IS ITS BILL?
Is it small and fine like a warbler's (1); stout and short like a seed-cracking sparrow's (2); dagger-shaped like a tern's (3); or hook-tipped like that of a bird of prey (4)?

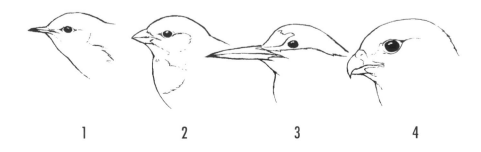

1	2	3	4

WHAT SHAPE IS ITS TAIL?

Is it deeply forked like a Barn Swallow's (1); square-tipped like a Cliff Swallow's (2); notched like a Tree Swallow's (3); rounded like a Blue Jay's (4); or pointed like a Mourning Dove's (5)?

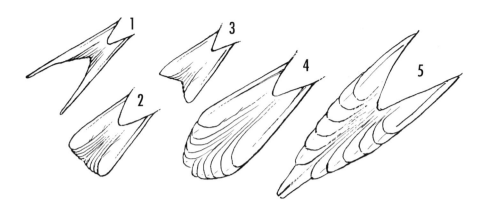

HOW DOES IT BEHAVE?

Does it cock its tail like a wren or hold it down like a flycatcher? Does it wag its tail? Does it sit erect on an open perch, dart after an insect, and return as a flycatcher does?

DOES IT CLIMB TREES?

If so, does it climb in spirals like a Creeper (left), in jerks like a woodpecker (center) using its tail as a brace, or does it go down headfirst like a nuthatch (right)?

3

HOW DOES IT FLY?

Does it undulate (dip up and down) like a Flicker (1)? Does it fly straight and fast like a Dove (2)? Does it hover like a Kingfisher (3)? Does it glide or soar?

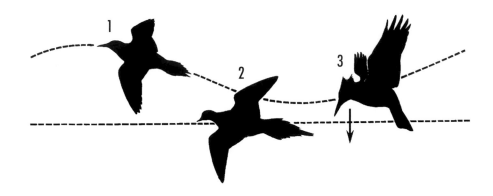

DOES IT SWIM?

Does it sit low in the water like a loon (1) or high like a gallinule (2)? If a duck, does it dive like a deepwater duck (3); or does it dabble and upend like a Mallard (4)?

DOES IT WADE?

Is it large and long-legged like a heron or small like a sandpiper? If one of the latter, does it probe the mud or pick at things? Does it teeter or bob?

WHAT ARE ITS FIELD MARKS?

Some birds can be identified by color alone, but most birds are not that easy. The most important aids are what we call field marks, which are, in effect, the "trademarks of nature." Note whether the breast is spotted as in the Wood Thrush (1); streaked as in the thrasher (2); or plain as in a cuckoo (3).

TAIL PATTERNS

Does the tail have a "flash pattern"—a white tip as in the Kingbird (1); white patches in the outer corners as in the towhee (2); or white sides as in the junco (3)?

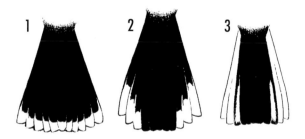

RUMP PATCHES

Does it have a light rump like a Cliff Swallow (1) or flicker (2)? The harrier, Yellow-rumped "Myrtle" Warbler, and many of the shorebirds also have distinctive rump patches.

5

EYE-STRIPES AND EYE-RINGS

Does the bird have a stripe above, through, or below the eye, or a combination of these stripes? Does it have a striped crown? A ring around the eye or "spectacles"? A "mustache" stripe? These details are important in many small songbirds.

WINGBARS

Do the wings have light wingbars or not? Their presence or absence is important in recognizing many warblers, vireos, and flycatchers. Wingbars may be single or double, bold or obscure.

WING PATTERNS

The basic wing patterns of ducks (shown below), shorebirds, and other water birds are very important. Notice whether the wings have patches (1) or stripes (2); are solidly colored (3) or have contrasting black tips (Snow Goose, etc.).

TOPOGRAPHY OF A BIRD

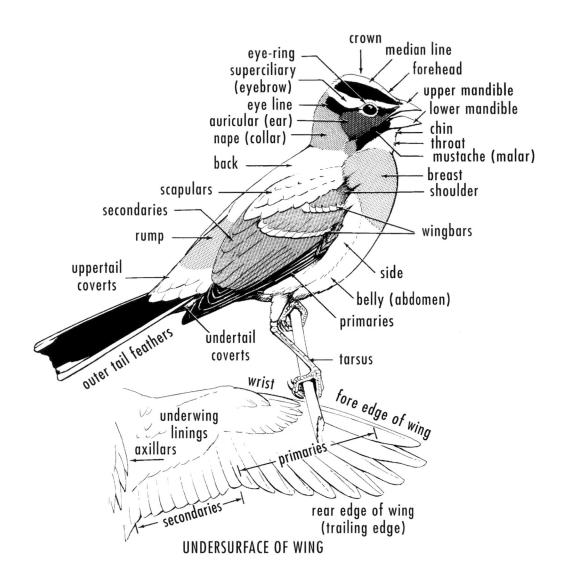

UNDERSURFACE OF WING

OTHER TERMS USED IN THIS BOOK

GENDER SYMBOLS: ♂ means male, ♀ means female. These symbols are used frequently on the plates, sparingly in the text.

ACCIDENTAL: In the area of this book, recorded fewer than 15 times; far out of range. On the state level only 1 or 2 records; might not be expected again.

CASUAL: Very few records, but might be expected again.

INTRODUCED: Not native; deliberately released.

EXOTIC: Not native; either released or escaped.

IN PART: A well-marked subspecies or form—part of a species.

CAUTION IN SIGHT RECORDS: Eighty years ago, prior to the Field Guide era, most ornithologists would not accept sight records of unusual birds unless they were made along the barrel of a shotgun. Today it is difficult to secure a collecting permit unless one is a professional or a student training to be one. Moreover, rarities may show up in parks, refuges, or on other lands where collecting is out of the question. There is no reason why we should not trust our increasingly educated eyes.

To validate the sight record of a very rare or accidental bird—a state "first," for example—the rule is that at least 2 competent observers should see the bird and document it in detail in their notes. A 35 mm camera equipped with a 400 mm lens is a useful tool for substantiating such sightings. Rarities are sometimes caught in mist nets by bird banders and can be photographed, hand-held, for the record. For this purpose a 50 mm close-up lens is best. Video cameras are also becoming popular.

There are some species—or plumages—that even the expert will hesitate to identify. And it is the mark of an expert to occasionally put a question mark after certain birds on his list: for example, accipiter, sp.?, or empidonax, sp.?, or "peep," sp.?, or immature Thayer's Gull? Do not be embarrassed if you cannot name *every* bird you see. Allan Phillips argued convincingly in *American Birds* (now known as *Field Notes*) that practically all of the Semipalmated Sandpipers so freely reported in winter on the southern coasts of the U.S. were really Western Sandpipers. It is extremely difficult to identify many individuals of these 2 species correctly in the winter.

BIRD NESTS: Most birders are not too skilled at finding nests. In most cases there would be an appalling gap between the number of species ticked off on their checklists and the number of nests they have discovered. To remedy this Hal Harrison, the premier nest photographer, prepared *A Field Guide to Birds' Nests*. This Field Guide will expand your ornithological expertise.

LARGE FORMAT EDITION
A FIELD GUIDE TO THE

BIRDS

EASTERN AND CENTRAL
NORTH AMERICA

Large aquatic diving birds. Must run over surface of water before taking off. Flight slow and with neck held lower than body. **FOOD:** Fish and crustaceans.

COMMON LOON *Gavia immer* 36"
VOICE: A haunting, laughing yodel.
HABITAT: Summer: lakes, tundra ponds; winter: coastal waters.
NOTES: Large nest of aquatic vegetation on shore edge. Cannot walk on land.

YELLOW-BILLED LOON *Gavia adamsii* 38"
VOICE: A hollow, deep yodel.
HABITAT: Tundra lakes, coastline.
NOTES: Extremely rare in East. No gray on bill and lower bill angled upward.

RED-THROATED LOON *Gavia stellata* 25"
VOICE: A ducklike quacking.
HABITAT: Tundra pools, coastal waters.
NOTES: Carries head on upward angle. Gray in winter compared to other loons.

PACIFIC LOON *Gavia pacifica* 26"
VOICE: A deep, growling *krrow*.
HABITAT: Tundra pools and coastal waters.
NOTES: Rare in East. Small Common and Red-throated Loons often mistaken for this species.

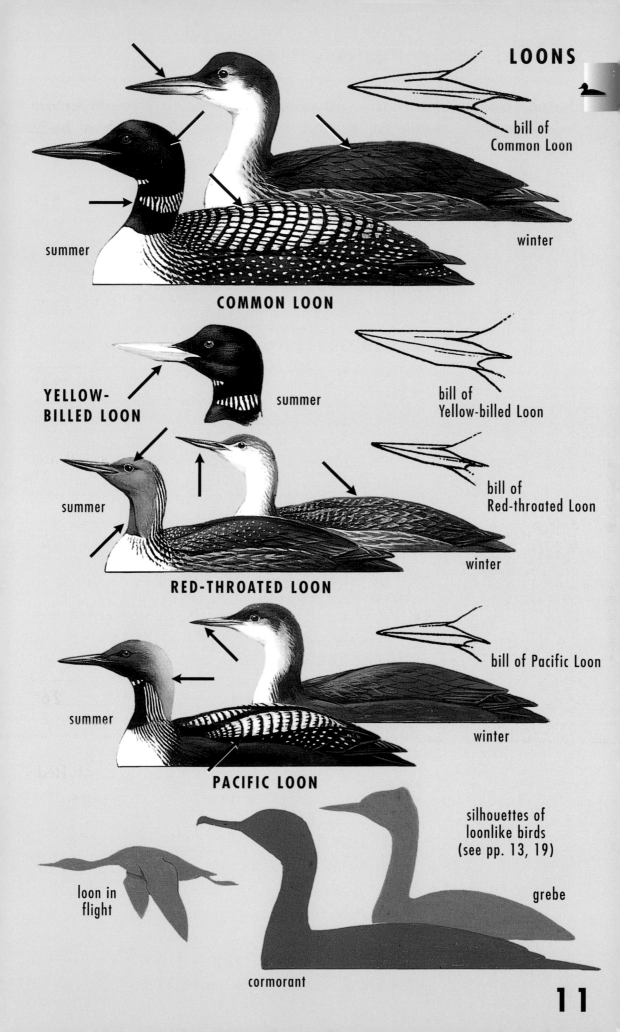

LOONS

bill of
Common Loon

summer

winter

COMMON LOON

**YELLOW-
BILLED LOON**

summer

bill of
Yellow-billed Loon

summer

summer

winter

bill of
Red-throated Loon

RED-THROATED LOON

bill of Pacific Loon

summer

winter

PACIFIC LOON

loon in
flight

silhouettes of
loonlike birds
(see pp. 13, 19)

grebe

cormorant

11

GREBES Family Podicipedidae

Small, trim aquatic birds with thin bills. Dive to feed. Must run across surface to take flight. **FOOD:** Fish, crustaceans, tadpoles, and insects.

HORNED GREBE *Podiceps auritus* to 15"
VOICE: A squeaky trill near nest; harsh *kerra*.
HABITAT: Lakes, ponds, and coastal waters.
NOTES: Common winter bird of East Coast waters and large inland lakes. Note pale neck and distinct cap.

EARED GREBE *Podiceps nigricollis* to 14"
VOICE: A froglike *poo-eep* or *kreeep*.
HABITAT: Prairie lakes and ponds, bays, coasts.
NOTES: Rare to East Coast waters in fall and winter. Note dark neck, cheeks, and peaked head.

PIED-BILLED GREBE
Podilymbus podiceps to 13"
VOICE: A cuckoolike *kuk-kuk-cow-cow-cow*.
HABITAT: Ponds, lakes, marshes, salt bays.
NOTES: Breeds locally and sparingly in East.

RED-NECKED GREBE
Podiceps grisegena to 18"
VOICE: A rapid *kik-kik-kik*.
HABITAT: Lakes, ponds; coastal in winter.
NOTES: Largest eastern grebe. Long neck and white chin patch, dull yellow at base of bill.

WESTERN GREBE
Aechmophorus occidentalis to 25"
VOICE: *Kirik*.
HABITAT: Reedy lakes, sloughs, bays in winter.
NOTES: Very rare in East. In near look-alike Clark's Grebe (*A. clarkii*) white of face extends above eye and bill is orange-yellow.

GREBES

lobed foot
of grebe

bill

winter

summer

HORNED GREBE

silhouette of
grebe in flight

bill

winter

summer

EARED GREBE

summer

winter

juvenile

downy
young

PIED-BILLED GREBE

summer

winter

immature

RED-NECKED GREBE

gallinule

coot

**CLARK'S
GREBE**

display

two other grebelike
birds (see p. 43)

WESTERN GREBE

13

Black and white sea birds that replace penguins in N. Hemisphere. Nest on sea cliffs, spend most of life on open waters of ocean. **FOOD:** Fish, crustaceans, mollusks, algae.

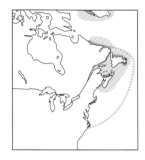

RAZORBILL *Alca torda*   to 18"
VOICE: A weak whistle or deep growling, at breeding sites.
HABITAT: Breeds on sea cliffs of N. Atlantic; spends rest of year on open ocean.
NOTES: Large bill is diagnostic. Often swims with tail cocked up.

THICK-BILLED MURRE *Uria lomvia*   to 19"
VOICE: Deep resonant growls, at breeding sites.
HABITAT: Breeds on islands and sea cliffs of N. Atlantic; spends rest of year on open ocean.
NOTES: Deeper black color above than Common Murre. Note white line near mouth gape.

COMMON MURRE *Uria aalge*   to 17"
VOICE: Deep rolling growls, at breeding sites.
HABITAT: Breeds on sea cliffs of N. Atlantic; spends rest of year on open ocean.
NOTES: More brownish color above than Thick-billed Murre.

ALCIDS (AUKS)

immature

summer

RAZORBILL   winter

winter

THICK-BILLED MURRE

summer

winter

COMMON MURRE

"Bridled" form   summer   THICK-BILLED MURRE

summer   summer

RAZORBILL
summer

1844

GREAT
AUK
extinct

15

DOVEKIE *Alle alle*   to 9"

VOICE: Shrill chattering, at breeding sites.

HABITAT: Offshore ocean waters. Returns to coastal sea cliffs to breed.

NOTES: A tiny sea bird that can occur by the thousands. Sometimes driven ashore and inland by storms.

BLACK GUILLEMOT *Cepphus grylle*   to 14"

VOICE: A very high-pitched, hissed *peeee* at breeding sites.

HABITAT: Inshore ocean waters. Breeds in openings of sea cliffs.

NOTES: The alcid most commonly seen from shore. White wing patch very distinctive. Reddish legs and feet.

ATLANTIC PUFFIN *Fratercula arctica* to 12"

VOICE: Growling at nest site.

HABITAT: Primarily offshore waters. Nests in turf burrows on coastal islands.

NOTES: With bright bill, best known of the alcids. Loses bright bill sheath for winter and moves out to sea.

winter

summer

DOVEKIE

summer

winter

summer

winter

BLACK GUILLEMOT

summer

summer

immature

ATLANTIC PUFFIN

winter

summer

BLACK GUILLEMOT
summer

adults
summer

DOVEKIE
summer

ATLANTIC PUFFIN

17

CORMORANTS Family Phalacrocoracidae

Large blackish water birds. Often stand erect on breakwaters and posts with wings in spread-eagle position. **FOOD:** Fish, crustaceans.

DOUBLE-CRESTED CORMORANT
Phalacrocorax auritus   **33"**
VOICE: Mainly piglike grunts at nesting colony.
HABITAT: Coasts, islands, bays, rivers, lakes.
NOTES: Migrates in large V's along coast. Breeds locally south of main breeding range.

GREAT CORMORANT
Phalacrocorax carbo   **37"**
VOICE: Grunts, growls, and brays at nest sites.
HABITAT: Mostly coastal; nests on sea cliffs.
NOTES: Its heavy bill, thick neck, white throat and flank patches (spring/summer) distinguish it from Double-crested. Immatures have dark breast, pale belly.

NEOTROPIC CORMORANT
Phalacrocorax brasilianus   **25"**
VOICE: Grunts and coos at nesting sites.
HABITAT: Tide waters, coastal lakes.
NOTES: Western Gulf of Mexico. Smaller head and neck than Double-crested. White border to yellow throat pouch.

DARTERS Family Anhingidae

Similar to cormorants but neck is snakier and bill is pointed. Silver wing and back pattern. One species in the U.S. **FOOD:** Fish and aquatic life.

ANHINGA *Anhinga anhinga*   **34"**
VOICE: A rattling, gargling *grrr, err, err, err.*
HABITAT: Cypress swamps, rivers, ponds.
NOTES: Often swims with only its snakelike head above water. Spreads wings and suns. Often soars high overhead in migration.

CORMORANTS

DOUBLE-
CRESTED
CORMORANT

immature

adult summer

adult

GREAT
CORMORANT

immature

DOUBLE-CRESTED
CORMORANT

adult

GREAT
CORMORANT

breeding

DOUBLE-
CRESTED
CORMORANT

breeding

NEOTROPIC
CORMORANT

breeding

DARTERS

♂

♀

♂

♀

swimming

ANHINGA

19

SWANS Subfamily Cygninae

Huge all-white aquatic birds. Larger, with longer necks than geese. Young are brown to off-white. **FOOD:** Aquatic plants, fruits, and seeds.

MUTE SWAN *Cygnus olor* 60"
VOICE: Grunts and wheezes.
HABITAT: Bays, lagoons, inlets, ponds, parks.
NOTES: Introduced from Eurasia. Male has larger black nasal knob than female. Can be threatening when protecting young.

TUNDRA SWAN *Cygnus columbianus* 53"
VOICE: A mellow, high-pitched cooing, especially while in flight: *woo-ho, woo-ho, woo-ho.*
HABITAT: Summer: tundra lakes; winter: bays, estuaries.
NOTES: Black bill has yellow spot in front of eye. Neck carried straight up—no curve.

GEESE Subfamily Anserinae

Large, heavy-bodied waterfowl. Call in flight. Most fly in V formation.

SNOW GOOSE *Chen caerulescens* to 38"
VOICE: A loud, nasal-like bark: *whouk.*
HABITAT: Tundra (summer); marshes, bays, grain fields, and ponds in winter and during migration.
NOTES: Head often stained reddish from feeding in iron-bound waters of north. Pink bill has distinct black grin patch.

ROSS'S GOOSE *Chen rossii* 23"
VOICE: High-pitched, doglike "yaps."
HABITAT: Tundra (summer); marshes, grain fields (winter).
NOTES: Rare in East. Note small size, petite bill with bluish base and without grin patch, steep forehead.

immature

adult

MUTE SWAN

adult

TUNDRA SWAN

immature

adult

MUTE SWAN

immature

adult

TUNDRA SWAN

immature

adult

ROSS'S GOOSE

SNOW GOOSE
white morph

ROSS'S

SNOW GOOSE

intergrade between
gray and white morphs

white morph

SNOW

21

SNOW GOOSE (Blue Goose)
Chen caerulescens   **to 28"**
VOICE: A loud, barking *whouk.*
HABITAT: Tundra (summer), marshes, ponds, grain fields, bays in winter.
NOTES: A color morph of the Lesser Snow Goose subspecies. Most winter on Gulf Coast.

GREATER WHITE-FRONTED GOOSE
Anser albifrons   **to 30"**
VOICE: A high-pitched tooting: *kah-lah-a-kuk.*
HABITAT: Tundra in summer; lakes, ponds, bays, grain fields in winter.
NOTES: Rare in East. Found in large goose flocks.

CANADA GOOSE
Branta canadensis   **25 to 43"**
VOICE: Musical and familiar honking.
HABITAT: Nonmigratory populations all year south to dash line. Wild: Canada and north in winter ponds, bays.
NOTES: Size is variable by subspecies. Nonmigratory birds becoming a nuisance in some places. Migrate in typical V's during migration.

BRANT *Branta bernicla*   **to 26"**
VOICE: A guttural, throaty *cr-r-r-uk* or *kurr-onk.*
HABITAT: Summer: Arctic coasts; winter: salt bays, and estuaries; scarce inland.
NOTES: Feeds on seaweeds along coast and marsh grasses in spring. Looks like small dark Canada Goose.

BARNACLE GOOSE *Branta leucopsis* **to 26"**
VOICE: A barking *yap.*
HABITAT: High Arctic, Greenland.
NOTES: Accidental or casual on East Coast. Ever-increasing sightings of birds mixed in with large goose flocks. Many individuals likely are escapes.

immature

adult

SNOW GOOSE
gray morph (Blue Goose)

adult

immature

adult

**GREATER WHITE-
FRONTED GOOSE**

adult

Richardson's
race

Lesser
race

Atlantic
race

CANADA GOOSE

**BARNACLE
GOOSE**

BRANT

GEESE AND SWANS IN FLIGHT

CANADA GOOSE *Branta canadensis*   p. 22
Light chest, black neck "stocking," white chinstrap.

BRANT *Branta bernicla*   p. 22
Small; black chest, head, and neck.

GREATER WHITE-FRONTED GOOSE *Anser albifrons*   p. 22
ADULT: Gray neck, black splotches on belly.
IMMATURE: Dusky, light bill, light feet.

SNOW GOOSE *Chen caerulescens*   p. 20
ADULT: White with black primaries.

SNOW GOOSE (GRAY MORPH) or BLUE GOOSE
Chen caerulescens **(IN PART)**   p. 22
ADULT: Dark body, white head.
IMMATURE: Dusky; dark bill, dark feet.

TUNDRA SWAN *Cygnus columbianus*   p. 20
Very long neck; plumage entirely white.

Many geese and swans fly in line or wedge formation.

CANADA
GOOSE

below

above

BRANT

adult

immature

GREATER WHITE-
FRONTED GOOSE

immature

SNOW
GOOSE
gray morph
(Blue Goose)

adult

SNOW
GOOSE

TUNDRA
SWAN

SNOW
GOOSE

adult

white morph

gray morph
(Blue Goose)

WHISTLING-DUCKS Subfamily Dendrocygninae

Long-legged, long-necked birds that roost and nest in trees.

FULVOUS WHISTLING-DUCK
Dendrocygna bicolor   to 21"
VOICE: A slurred whistle: *pee-whoo-ooo*.
HABITAT: Mainly coastal marshes and ponds.
NOTES: Gulf Coast. Recent invader of southern East Coast. Not a tree species like others in subfamily.

DABBLING DUCKS Subfamily Anatinae

Typical surface-feeding ducks. Can jump directly from surface of water into flight. Note colored patch (speculum) on rear of wing.

AMERICAN BLACK DUCK
Anas rubripes   to 25"
VOICE: Female: typical duck quack; male: low croak.
HABITAT: Marshes, bays, rivers, coasts, ponds.
NOTES: One of the most common ducks; often hybridizes with Mallard. Dark brown, not black!

MOTTLED DUCK *Anas fulvigula*   to 20"
A pale brown version of the American Black Duck with clear, pale throat. Confined to southern ponds, lakes, and backwaters.

GADWALL *Anas strepera*   to 23"
VOICE: Male: whistle, low *bek*; female: nasal quack.
HABITAT: Lakes, ponds, marshes.
NOTES: White wing patch. Black rump in male. Breeds locally within dash lines.

MALLARD *Anas platyrhynchos*   to 28"
VOICE: Male: low *yeeb*; female: loud quacking.
HABITAT: Marshes, woods, ponds, parks, lakes, bays.
NOTES: Perhaps the best-known duck. Common in parks. Male has an all-green head. Differentiate female from female American Black Duck by white edge to blue speculum and by bill color. Feral birds nest south to dash line.

FULVOUS WHISTLING-DUCK

sexes similar

MOTTLED DUCK

♂

♀

♂

AMERICAN BLACK DUCK

marsh ducks tip up

♂

♀

GADWALL

marsh ducks spring directly from the water

♀

♂

MALLARD

27

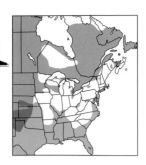

NORTHERN PINTAIL
Anas acuta   **to 30"**
VOICE: A 2-toned *prrip.*
HABITAT: Marshes, prairies, fresh ponds, lakes, bays.
NOTES: Head is small and rounded. Blue bill and pointed tail in both species, all plumages. Length includes long pin tail.

AMERICAN WIGEON *Anas americana* **to 20"**
VOICE: A high, whistled *whee whew.*
HABITAT: Marshes, lakes, bays, fields.
NOTES: Large white forewing patch; head is white or cream-colored. Often grazes on land.

EURASIAN WIGEON *Anas penelope*   **to 20"**
VOICE: A whistled *wheeoo.*
HABITAT: Ponds, bays, coastal waters.
NOTES: Breeds in Iceland. Rare winter visitor. Found in American Wigeon flocks. Male: rusty head, buffy crown stripe, and gray sides. Female: tawnier head than American Wigeon's.

WOOD DUCK *Aix sponsa*   **to 20"**
VOICE: Male: *creek creek* or ascending *freeee;* female: *whoo-eek.*
HABITAT: Wooded swamps, ponds, river edges.
NOTES: Male's colors spectacular. Female has white tear-shaped eye patch. Note long square tail in flight. Nests in nest boxes and tree hollows, up to a mile from water.

POSTURES OF DUCKS

The shape of a duck often gives a clue to the group it belongs to. The silhouette of a marsh duck reveals that the legs are well forward, which allows the duck to forage on land. A bay duck with legs far to the rear is built for diving and is awkward on land.

♀

♂

NORTHERN PINTAIL

♂

♀

AMERICAN WIGEON

♂

♀

EURASIAN WIGEON

♀

♂ in eclipse (autumn)

♂

WOOD DUCK

POSTURES OF DUCKS ON LAND

Marsh and
Pond Ducks
(dabblers)

Sea and
Bay Ducks
(divers)

Mergansers
(divers)

Ruddy
Duck
(diver)

Whistling
Ducks
(dabblers) **29**

NORTHERN SHOVELER
Anas clypeata   **to 20"**
VOICE: Male: low *took, took*; female: low *quack*.
HABITAT: Freshwater marshes, ponds, sloughs, salt bays in winter.
NOTES: Distinct spatulate bill for filter feeding. Powder blue forewing in flight. Breeds locally south to dash line.

BLUE-WINGED TEAL *Anas discors*   **to 16"**
VOICE: Male: peeping notes; female: a low *quack*.
HABITAT: Fresh ponds and marshes.
NOTES: Small size distinguishes teal. Blue forewing separates it from Green-winged. Breeds locally south to dash line.

GREEN-WINGED TEAL *Anas crecca*   **to 14"**
VOICE: Male: a short whistle and froglike peeping; female: a sharp, crisp *quack*.
HABITAT: Marshes, river, bays.
NOTES: 2 distinct subspecies: American race has a vertical white shoulder mark; Eurasian race (a rare visitor from Iceland) has a horizontal mark that runs along top of folded wing.

Accidental and Escaped

CINNAMON TEAL *Anas cyanoptera*   **to 17"**
Found in marshes, prairie ponds, and coastal estuaries. A casual species from the prairie states and west to the East Coast. Breeding male is the only rich chestnut duck with blue forewings.

MUSCOVY DUCK *Cairina moschata*   **to 32"**
This species occurs in the wild only along the Lower Rio Grande River of Texas. It is a common domesticated species and is often released in ponds and lakes throughout the East. Very dimorphic in color patterns. Red facial knobs and bare facial skin may be absent in female. Almost gooselike.

♂ ♀

NORTHERN SHOVELER

♂ ♀

BLUE-WINGED TEAL

♂ ♀

GREEN-WINGED TEAL
American race

♂

GREEN-WINGED TEAL
Eurasian race

♂

CINNAMON TEAL

ancestral form

domestic variation

MUSCOVY DUCK
(a frequent escape from captivity)

Also known as bay ducks, they are found on lakes and rivers and breed in marshes. All dive. To take flight they must run over surface of water. Scoters often swim with tail cocked up. **FOOD:** Small aquatic animals and plants. Seagoing species eat mollusks and crustaceans.

WHITE-WINGED SCOTER

Melanitta fusca   to 21"

VOICE: A low grunting; wings whistle in flight.

HABITAT: Summer: lakes; winter: bays and ocean.

NOTES: Distinctly larger than other scoters. White wing patch can be difficult to see when bird is floating, so look for white eye mark and feathering on base of bill on male.

SURF SCOTER

Melanitta perspicillata   to 19"

VOICE: Croaks or grunts.

HABITAT: Summer: Arctic lakes and tundra ponds; winter: ocean coasts, bays.

NOTES: Female's head spots are shaped slightly differently than those of female White-winged Scoter. Scoters often form massive rafts of thousands offshore.

BLACK SCOTER   *Melanitta nigra*   to 18½"

VOICE: Male: melodious cooing; female: low growls.

HABITAT: Summer: coastal tundra ponds; winter: sea coasts, large bays. All three species occur regularly on Great Lakes. Watch for migrants on interior lakes.

NOTES: Knob on bill becomes orange-yellow in spring. Female's pale cheeks contrast with her dark cap.

SEA DUCKS (SCOTERS)
(Divers)

scoters fly in line
or V formation

White-winged

Surf

Black

immature

♂

♀

WHITE-WINGED SCOTER

♂

♀

immature

SURF SCOTER

♂

♀

inset
**LABRADOR
DUCK**
extinct

1878

♂

BLACK SCOTER

diving ducks (sea ducks and bay ducks)
raft on water, skitter when taking wing

OLDSQUAW *Clangula hyemalis*
male 21" (with tail), female 16"
VOICE: A musical, babbling *ow-owdle-ow*.
HABITAT: Summer: tundra lakes and pools; winter: ocean, shoreline, large bays and lakes.
NOTES: Distinct black wings. Male has pale color on bill; female's bill is entirely dark.

HARLEQUIN DUCK
Histrionicus histrionicus **to 18"**
VOICE: Male: squeaks and *qwa-qwa*; female: *ek, ek*.
HABITAT: Summer: turbulent mountain streams; winter: rocky coastal waters in rock wash zones.
NOTES: Small active ducks that bob like corks. Female has small bill, three white head spots.

KING EIDER *Somateria spectabilis* **to 24"**
VOICE: Male: low, crooning *carooo*; female grunts.
HABITAT: Summer: high Arctic pools; winter: coasts and open ocean.
NOTES: Young males: white chest, dark head and body. Females: golden or sandy brown. A few on Great Lakes.

COMMON EIDER
Somateria mollissima **to 27"**
VOICE: Male: cooing *ow-ooo-urr*; female: *kor-r-r*.
HABITAT: Summer: coasts, islands, tundra ponds; winter: rocky coasts, shoals—often in large rafts.
NOTES: Slope of forehead separates from King. Female is true brown.

♂ summer   ♂ winter   **OLDSQUAW**   ♀ summer

♀ winter

HARLEQUIN DUCK

♂   ♀

♂ immature

♂

bill of ♀ King Eider

KING EIDER

♂ immature

♂   ♀

bill of ♀ Common Eider

COMMON EIDER

King Eider

Common Eider

eiders raft in large flocks

35

CANVASBACK *Aythya valisineria*   **to 24"**
VOICE: Male: low croaks; female: *quack*.
HABITAT: Fresh marshes, lakes; winter: lakes, salt bays, estuaries.
NOTES: Slope of forehead to bill is distinct, back is pale. Fast-flying.

REDHEAD *Aythya americana*   **to 23"**
VOICE: Male: cooing *errr*, also harsh, catlike *meow*; female: an explosive *squak*.
HABITAT: Summer: fresh marshes; winter: lakes, bays.
NOTES: Dark back and steep forehead. Pale ring atn-ear bill tip.

RING-NECKED DUCK *Aythya collaris* **to 18"**
VOICE: A low *purr*; low whistle when courting.
HABITAT: Wooded lakes, rivers, ponds, river edges.
NOTES: Distinguish female from Redhead and Scaup by bill pattern, head shape, and face pattern. Neck collar is difficult to see.

LESSER SCAUP *Aythya affinis*   **to 18"**
VOICE: Low *quack* or *scaup*.
HABITAT: Summer: marsh ponds; winter: lakes, bays, nearshore ocean waters.
NOTES: Only its peaked head and white on interior wing help separate it from Greater Scaup. Dusk gray sides.

GREATER SCAUP *Aythya marila*   **to 20"**
VOICE: A squabbling *quack* or *scaup*.
HABITAT: Summer: tundra pools; winter: lakes, rivers, bays, estuaries, and nearshore ocean waters.
NOTES: Greater has a rounded head and slightly wider bill than Lesser. Often form massive rafts. Tufted Duck (*A. fuligula*), a stray from Iceland and Europe, sometimes seen with scaup flocks.

CANVASBACK

♀

♂

diving ducks run and patter

REDHEAD

♀

♂

TUFTED DUCK
(accidental)

♂

RING-NECKED DUCK

♀

♂

Lesser

Greater

Greater

Lesser

♀

♂

♂

GREATER SCAUP

LESSER SCAUP

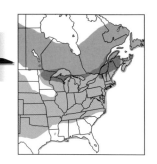

COMMON GOLDENEYE
Bucephala clangula   **to 20"**
VOICE: Male gives nasal whistle in courtship; female: harsh *quack*. Wings whistle in flight.
HABITAT: Summer: forested lakes and rivers; winter: lakes, rivers, salt bays, nearshore ocean waters.
NOTES: Male's green head with round spot distinguishes it from Barrow's with purple head and crescent spot. Form large offshore rafts.

BARROW'S GOLDENEYE
Bucephala islandica   **21"**
VOICE: Male: mewing cry; female: hoarse *quack*.
HABITAT: Summer: wooded lakes, beaver ponds; winter: large inland rivers, coastal waters.
NOTES: Male shows black smudge at shoulder. Barrow's has steeper forehead and stubbier bill than Common Goldeneye.

BUFFLEHEAD *Bucephala albeola*   **to 15"**
VOICE: Male: hoarse rolling note; female: *quack*.
HABITAT: Lakes, ponds, rivers; salt bays in winter.
NOTES: Small ducks. From a distance males look all white. Small white patch behind eye in female is distinct.

STIFF-TAILED DUCKS Subfamily Oxyurinae

Small; nearly helpless on land. Spike tail carried upright. Broad bill for filtering. **FOOD:** Aquatic life, insects, water plants.

RUDDY DUCK *Oxyura jamaicensis*   **to 16"**
VOICE: Male: a sputtering *chick-ik-ik-ik-k-k-kwrr*.
HABITAT: Fresh ponds, marshes, prairie pools; winter: salt bays, rivers.
NOTES: Small duck with broad bill. Buzzy flight. Females have white cheeks with a line across them.

♂ **COMMON GOLDENEYE** ♀

♀ breeding

♀ winter

♂ **BARROW'S GOLDENEYE**

♂ **BUFFLEHEAD** ♀

♂ summer **RUDDY DUCK** ♂ winter ♀

Diving ducks with spikelike bills, saw-edged mandibles. In flight, the head is held below plane of back. **FOOD:** Chiefly small fish.

COMMON MERGANSER
Mergus merganser   **to 27"**

VOICE: Male: staccato croaks; female: low *karr.*

HABITAT: Summer: wooded lakes and ponds; winter: open lakes and rivers, rarely estuaries.

NOTES: Prefers fresh water but will use brackish water. Rarely seen on open coastal waters. Sharp contrast of rusty head to whitish chest and distinct whitish throat in female.

RED-BREASTED MERGANSER
Mergus serrator   **to 26"**

VOICE: Male: guttural croaks; female: *karr.*

HABITAT: Summer: lakes, rivers; winter: open lakes, coastal bays, and coastal waters.

NOTES: Much more of a saltwater bird in winter than the Common. Distinct crest in both male and female. No distinct contrast between rust of head and gray chest in female.

HOODED MERGANSER
Lophodytes cucullatus   **to 19"**

VOICE: Male: low *karr-croo,* grunts; female: croaks.

HABITAT: Summer: wooded lakes, ponds, rivers; winter: open fresh water, estuaries.

NOTES: Small. Often keeps crest tucked on back. Fans crest during courtship. Female's subtle crest rusty compared to grayish body. Nests readily in boxes. Breeds locally and irregularly within dash lines.

MERGANSERS
(Divers)

mergansers fly with bill, head, body, and tail on the same horizontal axis

saw-edged mandibles of merganser

♂ ♀

COMMON MERGANSER

♂ ♀

RED-BREASTED MERGANSER

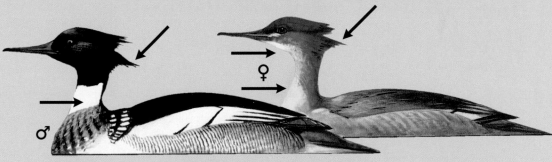

♂ crest up   ♂ crest down   ♀   immature

HOODED MERGANSER

Common   Red-breasted   Hooded

41

DUCKLIKE SWIMMERS (COOTS, MOORHENS, GALLINULES) Family Rallidae (in part)

Coots, moorhens, and gallinules belong to the same family as rails (covered on p. 90) but are included here based on their ducklike appearance. Coots dive, rails do not. **FOOD:** Aquatic insects, plants.

AMERICAN COOT *Fulica americana* **to 16"**
VOICE: A grating *kuk-kuk-kuk-ka-ka-ka-ka*, also a measured *ka-ka-ka*.
HABITAT: Ponds, lakes, marshes, cattail edges, fields, park ponds, golf courses.
NOTES: White bill, no side stripe. Bobs head when it swims. Often pilfers food from other diving birds. Feet have lobes, not webs.

COMMON MOORHEN
Gallinula chloropus **to 13"**
VOICE: Grunts, groans, cackles, a distinct froglike *kup*, and henlike notes.
HABITAT: Fresh marshes, reedy ponds.
NOTES: Red forehead shield. White side stripe even in immatures. Bobs head as it swims. Long toes aid in walking on vegetation.

PURPLE GALLINULE
Porphyrula martinica **to 13"**
VOICE: A henlike cackling: *kek-kek-kek*; guttural groans and croaks.
HABITAT: Freshwater swamps, marshes, ponds.
NOTES: Powder blue forehead shield above red bill. Iridescent violet purple. Extremely long toes for walking on lily pads. May show up well north of its normal range.

lobed foot of coot

coots skitter on takeoff

immature

coot chick

adult

adult

AMERICAN COOT

moorhen chick

immature

COMMON MOORHEN

adult

COMMON MOORHEN

immature

adult

PURPLE GALLINULE

adult

PURPLE GALLINULE

43

FLIGHT PATTERNS OF DABBLING DUCKS

NOTE: Males are described below. Females are somewhat similar.

NORTHERN PINTAIL (Sprig*) *Anas acuta* p. 28
OVERHEAD: Needle tail, white breast, thin neck.
TOPSIDE: Needle tail, neck stripe, one white border on rear edge of wing (speculum).

WOOD DUCK *Aix sponsa* p. 28
OVERHEAD: White belly, dusky wings, long square tail.
TOPSIDE: Stocky; long dark tail, white border on dark wing.

AMERICAN WIGEON (Baldpate*) *Anas americana* p. 28
OVERHEAD: White belly.
TOPSIDE: Large white shoulder patch.

NORTHERN SHOVELER (Spoonbill*) *Anas clypeata* p. 30
OVERHEAD: Dark belly, white chest, spoon bill.
TOPSIDE: Large bluish shoulder patch, spoon bill.

GADWALL *Anas strepera* p. 26
OVERHEAD: White belly.
TOPSIDE: White patch on rear edge of wing (speculum).

GREEN-WINGED TEAL *Anas crecca* p. 30
OVERHEAD: Small (teal-sized); light belly, dark head.
TOPSIDE: Small, dark-winged; green speculum.

BLUE-WINGED TEAL *Anas discors* p. 30
OVERHEAD: Small (teal-sized); dark belly.
TOPSIDE: Small; large bluish shoulder patch.

---

* Commonly used by duck hunters.

Wing of a marsh duck showing the iridescent speculum

DABBLING DUCKS   IN FLIGHT   Overhead

NORTHERN PINTAIL ♀   ♂

WOOD DUCK ♂   ♀

AMERICAN WIGEON ♀   ♂

NORTHERN SHOVELER ♀   ♂

GADWALL ♀   ♂

GREEN-WINGED TEAL ♀   ♂

BLUE-WINGED TEAL ♀   ♂

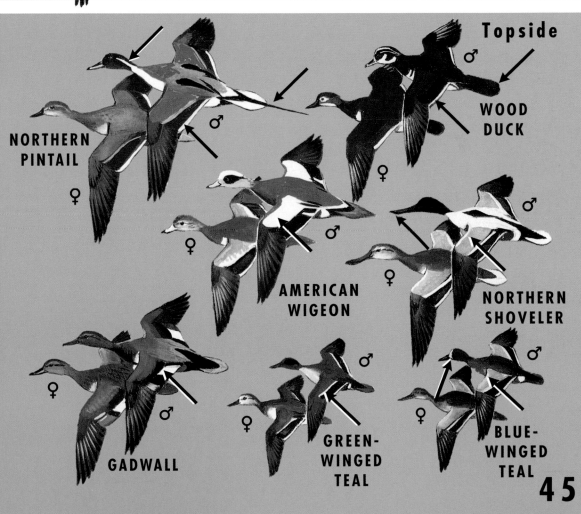

Topside

NORTHERN PINTAIL ♂   ♀

WOOD DUCK ♂   ♀

AMERICAN WIGEON ♀   ♂

NORTHERN SHOVELER ♀   ♂

GADWALL ♀   ♂

GREEN-WINGED TEAL ♀   ♂

BLUE-WINGED TEAL ♀   ♂

45

FLIGHT PATTERNS OF DABBLING DUCKS AND MERGANSERS

NOTE: Males are described below. Females are somewhat similar. Mergansers have a distinctive flight silhouette: the bill, head, neck, body, and tail are on a horizontal axis.

MALLARD *Anas platyrhynchos* p. 26
OVERHEAD: Dark chest, light belly, white neck-ring.
TOPSIDE: Dark head, neck-ring, 2 white borders on speculum.

AMERICAN BLACK DUCK *Anas rubripes* p. 26
OVERHEAD: Dusky body, white wing linings.
TOPSIDE: Dusky body, paler head.

FULVOUS WHISTLING-DUCK *Dendrocygna bicolor* p. 26
OVERHEAD: Tawny with blackish wing linings.
TOPSIDE: Dark unpatterned wings, white ring on rump.

COMMON MERGANSER* *Mergus merganser* p. 40
OVERHEAD: Merganser shape; black head, white body, white wing linings.
TOPSIDE: Merganser shape; white chest, large wing patches.

RED-BREASTED MERGANSER* *Mergus serrator* p. 40
OVERHEAD: Merganser shape; dark chest band.
TOPSIDE: Merganser shape; dark chest, large wing patches.

HOODED MERGANSER* *Lophodytes cucullatus* p. 40
OVERHEAD: Merganser shape; dusky wing linings.
TOPSIDE: Merganser shape; small wing patches.

---

* The duck hunter often calls mergansers "sheldrakes" or "sawbills."

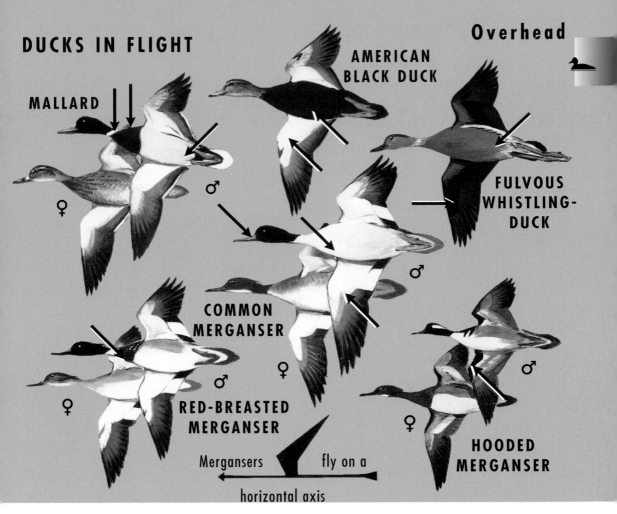

DUCKS IN FLIGHT

Overhead

MALLARD

AMERICAN
BLACK DUCK

♂

♀

FULVOUS
WHISTLING-
DUCK

COMMON
MERGANSER

♂

♀

♀

RED-BREASTED
MERGANSER

HOODED
MERGANSER

♂

♀

Mergansers   fly on a

horizontal axis

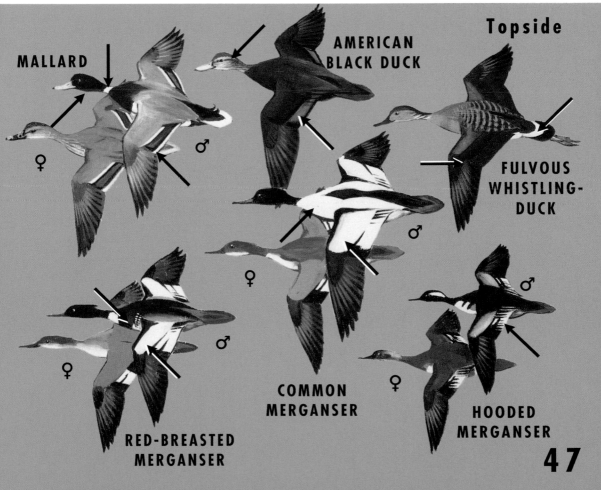

MALLARD

AMERICAN
BLACK DUCK

Topside

♂

♀

FULVOUS
WHISTLING-
DUCK

♂

♂

♀

RED-BREASTED
MERGANSER

♀

COMMON
MERGANSER

HOODED
MERGANSER

47

FLIGHT PATTERNS OF SEA DUCKS

NOTE: Only males are described below.

OLDSQUAW *Clangula hyemalis* p. 34
OVERHEAD: Dark unpatterned wings, white belly.
TOPSIDE: Dark unpatterned wings, much white on body.

HARLEQUIN DUCK *Histrionicus histrionicus* p. 34
OVERHEAD: Solid dark below, white head spots, small bill.
TOPSIDE: Dark with white marks, small bill.

SURF SCOTER* *Melanitta perspicillata* p. 32
OVERHEAD: Dark body, white head patches (not readily visible from below).
TOPSIDE: Dark body, white head patches. Immature and female have pale belly.

BLACK SCOTER* *Melanitta nigra* p. 32
OVERHEAD: Black plumage, paler flight feathers.
TOPSIDE: All-black plumage. Immature and female dark, with pale belly.

WHITE-WINGED SCOTER* *Melanitta fusca* p. 32
OVERHEAD: Black body, white wing patches.
TOPSIDE: Black body, white wing patches. Immature and female dark, with pale belly.

COMMON EIDER *Somateria mollissima* p. 34
TOPSIDE: Adult male has white back, white forewing.

KING EIDER *Somateria spectabilis* p. 34
TOPSIDE: Adult male has whitish foreparts, black rear parts.

*The three scoters are often called "coots" by duck hunters.

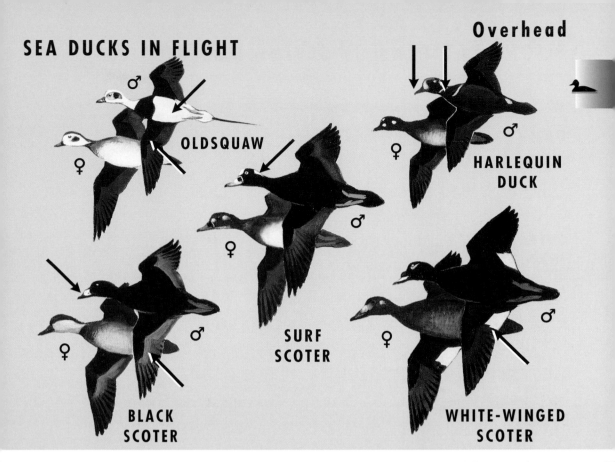

SEA DUCKS IN FLIGHT

Overhead

OLDSQUAW ♂ ♀

HARLEQUIN DUCK ♂ ♀

SURF SCOTER ♂ ♀

BLACK SCOTER ♂ ♀

WHITE-WINGED SCOTER ♂

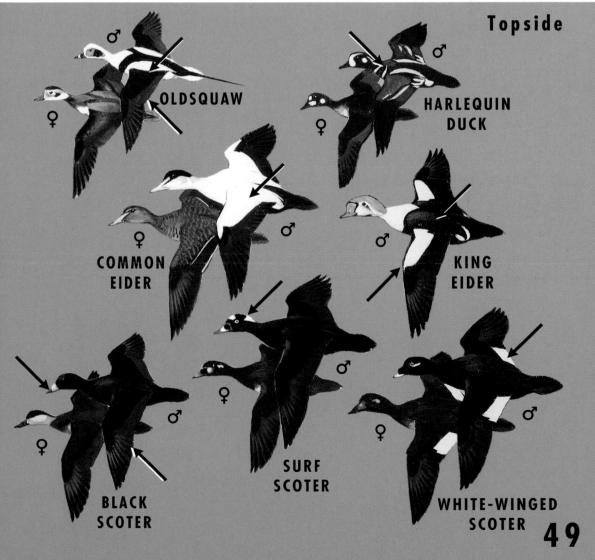

Topside

OLDSQUAW ♂ ♀

HARLEQUIN DUCK ♂ ♀

COMMON EIDER ♂ ♀

KING EIDER ♂

BLACK SCOTER ♂ ♀

SURF SCOTER ♂ ♀

WHITE-WINGED SCOTER ♂ ♀

49

FLIGHT PATTERNS OF DIVING DUCKS

NOTE: Only males are described below. The first 5 have black chests.

CANVASBACK *Aythya valisineria*   p. 36
OVERHEAD: Black chest, long profile.
TOPSIDE: White back, long profile.

REDHEAD *Aythya americana*   p. 36
OVERHEAD: Black chest, roundish rufous head.
TOPSIDE: Gray back, broad gray wing stripe.

RING-NECKED DUCK *Aythya collaris*   p. 36
OVERHEAD: Nearly impossible to tell from Scaup overhead.
TOPSIDE: Black back, broad gray wing stripe.

GREATER SCAUP (Bluebill*) *Aythya marila*   p. 36
OVERHEAD: Black chest, white stripe showing through wing.
TOPSIDE: Broad white wing stripe (extending onto primaries).

LESSER SCAUP (Bluebill*) *Aythya affinis*   p. 36
TOPSIDE: Wing stripe shorter than that of Greater Scaup.

COMMON GOLDENEYE (Whistler*) *Bucephala clangula*   p. 38
OVERHEAD: Blackish wing linings, white wing patches.
TOPSIDE: Large white wing-square, short neck, black head.

RUDDY DUCK *Oxyura jamaicensis*   p. 38
OVERHEAD: Stubby; white face, dark chest.
TOPSIDE: Small; dark with white cheeks.

BUFFLEHEAD (Butterball*) *Bucephala albeola*   p. 38
OVERHEAD: Looks like small goldeneye; note head patch.
TOPSIDE: Small; large wing patches, white head patch.

* Commonly used by duck hunters.

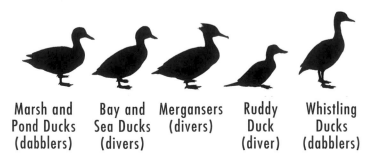

| Marsh and Pond Ducks (dabblers) | Bay and Sea Ducks (divers) | Mergansers (divers) | Ruddy Duck (diver) | Whistling Ducks (dabblers) |

POSTURES OF DUCKS ON LAND

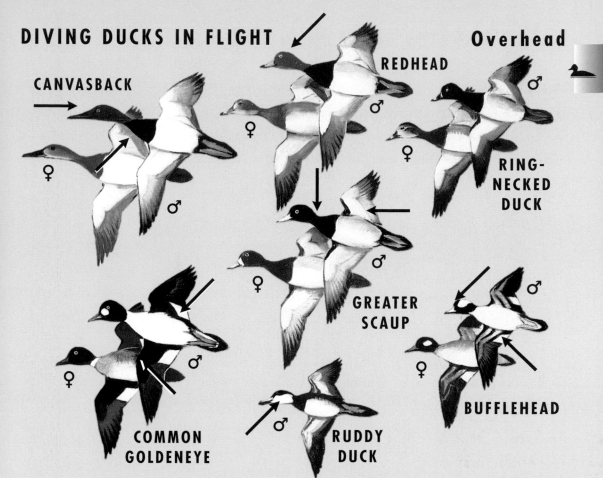

DIVING DUCKS IN FLIGHT

Overhead

CANVASBACK

REDHEAD

♂

♀

♀

♂

RING-
NECKED
DUCK

♀

GREATER
SCAUP

♂

BUFFLEHEAD

♀

♂

♀

COMMON
GOLDENEYE

♂

RUDDY
DUCK

Topside

CANVASBACK

REDHEAD

♂

♀

♂

♀

♂

RING-
NECKED
DUCK

GREATER
SCAUP

♂

below:
wing of
Lesser
Scaup

♂

♀

♀

COMMON
GOLDENEYE

♂

RUDDY
DUCK

BUFFLEHEAD

51

SHEARWATERS, etc. Family Procellariidae

Gull-like birds of the open sea with stiff, gliding flight low over the waves. They have narrow wings, short tails, and bills with external, tubelike nostrils. **FOOD:** Fish, squid, crustaceans, ship refuse. Most species on this page are nonbreeding summer visitors.

CORY'S SHEARWATER
Calonectris diomedea   **to 21"**
VOICE: Guttural croaks and moaning while nesting.
HABITAT: Oceanic. East Coast in summer and fall.
NOTES: Largest East Coast shearwater, gray-brown. Note pale bill. Flight is several flaps, then glide.

GREATER SHEARWATER
Puffinus gravis   **to 19"**
VOICE: Screeches during food fights.
HABITAT: Oceanic. Off East Coast spring to fall.
NOTES: Dark cap and distinct white rump patch. Follows boats; can appear in large numbers.

SOOTY SHEARWATER
Puffinus griseus   **to 17"**
VOICE: Grunts and groans on breeding islands.
HABITAT: Oceanic. To N. Atlantic in summer.
NOTES: Gray-brown with pale to whitish underwings.

MANX SHEARWATER
Puffinus puffinus   **13"**
VOICE: Moans and squeaks at nest site.
HABITAT: Oceanic. Summer visitor along North Atlantic coast.
NOTES: Small, black and white. Flight with long glides and roller-coaster glides.

AUDUBON'S SHEARWATER
Puffinus lherminieri   **to 12"**
VOICE: Groans on breeding grounds.
HABITAT: Oceanic. North regularly to North Carolina, in warm water with scattered sightings farther north.
NOTES: Dark undertail and browner cast than the Manx. Rapid, fluttering wingbeats, less gliding.

SHEARWATERS

GREATER
SHEARWATER

CORY'S
SHEARWATER

SOOTY
SHEARWATER

MANX
SHEARWATER

tubed bill
of shearwater

AUDUBON'S
SHEARWATER

53

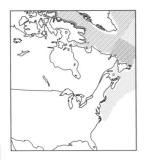

NORTHERN FULMAR
Fulmarus glacialis **to 18"**
VOICE: A hoarse *ag-ag-ag-arrr* or *ek-ek-ek*.
HABITAT: N. Hemisphere oceans. Nests on sea cliffs.
NOTES: Bulky; bull-necked, bulging forehead. In dark form it is brownish to gray. Flies like shearwater.

RARE PTERODROMA PETRELS
(Large, shearwater-like)

BLACK-CAPPED PETREL *Pterodroma hasitata* 16"
Small with white forehead and white rump patch like a Greater Shearwater. White below. Regular in Gulf Stream north to off North Carolina.

BERMUDA PETREL (Cahow) *Pterodroma cahow* 15"
Looks like a Black-capped Petrel, but nape of neck is dark. Pale rump band is narrow. Breeds only on Bermuda. A few records off North Carolina.

STORM-PETRELS Family Hydrobatidae

Little dark birds that flit over the ocean. Can appear in massive numbers. Nest on sea islands—come and go at night. **FOOD:** Plankton, crustaceans, some small fish.

WILSON'S STORM-PETREL
Oceanites oceanicus **7"**
VOICE: Twitters and chips at nest site.
HABITAT: Ocean wanderer; breeds in Antarctic. Ranges to Northeast coast in summer.
NOTES: One of the most common birds in the world! Drifts over surface like a swallow pausing to flutter and patter with feet. White rump, tail squared off.

LEACH'S STORM-PETREL
Oceanodroma leucorhoa **8"**
VOICE: Falsetto crooning notes at nest.
HABITAT: Oceanic waters.
NOTES: White rump, forked tail end. Flight is erratic with leaps and bounds. Does not follow ships.

PETRELS

tubed bill of fulmar, typical of petrels

light form

NORTHERN FULMAR

light form

dark form

NORTHERN FULMAR

BLACK-CAPPED PETREL

BERMUDA PETREL (Cahow) (upper)

BERMUDA PETREL (Cahow) (under)

STORM PETRELS

WILSON'S STORM-PETREL

LEACH'S STORM-PETREL

tubed bill of storm-petrel

PELICANS Family Pelecanidae

Huge water birds with long flat bills and great throat pouches. Sexes alike. Fly in lines or V's. In flight, head is hunched back on shoulders. **FOOD:** Mainly fish and crustaceans.

AMERICAN WHITE PELICAN
Pelecanus erythrorhynchos
to 62" (wingspan to 9½')
VOICE: Usually silent. Low groans at colony.
HABITAT: Lakes, marshes, salt bays.
NOTES: Does not plunge dive like Brown Pelican. Often flies high in V's or lines. Bill ridge known only in breeding season. Rare to East Coast.

BROWN PELICAN *Pelecanus occidentalis*
to 50" (wingspan to 6½')
VOICE: Usually silent; croaks when disturbed.
HABITAT: Salt bays, beaches, shoreline, marinas.
NOTES: Plunge dives with bill outstretched for food. Lines of birds skim over the water along coast.

FRIGATEBIRDS Family Fregatidae

Dark tropical sea birds with extremely long, angled wings and deeply forked tail. Cannot swim. **FOOD:** Fish, jellyfish, squid, young birds. Pirates fish from other birds.

MAGNIFICENT FRIGATEBIRD
Fregata magnificens
to 41" (wingspan 7 to 8')
VOICE: Guttural gargle at breeding colonies.
HABITAT: Oceanic coasts and islands.
NOTES: Soars with ease high overhead. Piratic, stealing food from other seabirds. Plucks food out of water with its long bill. Male displays with vermilion gular pouch. Female is white-chested; immature is white-headed. Summer visitor within dash line.

PELICANS,
FRIGATES

adult
breeding

immature

adults

adult
winter

AMERICAN
WHITE PELICAN

adults
summer

immature

BROWN
PELICAN

♂

♀

♂

immature

MAGNIFICENT
FRIGATEBIRD
♂ (left) in display

57

GANNETS, BOOBIES Family Sulidae

Sea birds with large tapering bills and pointed tails. Plunge dive to feed. Sexes alike. Young dark. FOOD: Fish, squid.

NORTHERN GANNET *Morus bassanus* to 38"
VOICE: A barking *arrah* at colony.
HABITAT: Oceanic. Breeds on northern sea cliffs.
NOTES: First-year birds flecked with spots, second-year birds piebald. Adults are white with black wingtips. Migrates in large groups.

MASKED BOOBY *Sula dactylatra* to 27"
VOICE: Guttural growls at colony.
HABITAT: Tropical oceans and islands. Wanders to Gulf Coast and up Gulf Stream; regular at Dry Tortugas.
NOTES: Looks like small gannet but with black tail and black trailing edge to wings.

BROWN BOOBY *Sula leucogaster* to 30"
VOICE: Croaking and guttural sounds at colony.
HABITAT: Tropical oceans: Gulf Coast; regular to Dry Tortugas area; rare in Gulf Stream farther north.
NOTES: All-brown head, white belly. Immature all-brown. Rest on marker buoys and fishing ship riggings.

TROPICBIRDS Family Phaethontidae

Similar to large terns (no relation); 2 elongate central tail feathers. Dive headfirst for food. FOOD: Squid, crustaceans.

WHITE-TAILED TROPICBIRD
Phaethon lepturus
32" (with 16" streamer tail feathers)
VOICE: A harsh, ternlike scream.
HABITAT: Pantropical oceans.
NOTES: Breeds in Bermuda, wanders Gulf Stream north, and south off southern Florida. Bold back pattern. Immature has yellow bill, adult orange.

GANNETS, BOOBIES, TROPICBIRDS

1st year

adults

NORTHERN GANNET

changing

diving

MASKED BOOBY

adults

immature

BROWN BOOBY

adult

adults

below

above

immature

immature

adult

WHITE-TAILED TROPICBIRD

59

JAEGERS, SKUAS Family Stercorariidae

Dark, falconlike sea birds. Ocean hunters. Powerful flight. Most jaegers migrate offshore. Harass sea birds to disgorge food. All but South Polar Skua nest in Arctic. Young are difficult to identify. **FOOD:** Arctic: lemmings, eggs, young birds. Ocean: food from other birds or water surface. **VOICE:** As a group, loud, harsh, strident scolds and screams.

GREAT SKUA *Catharacta skua*   to 24"
HABITAT: Ocean wanderer. Breeds in northern and Arctic waters. Winters off New England states.
NOTES: Massive; dark brown, no pale nape. Powerful wing strokes. No tail points like smaller jaegers.

SOUTH POLAR SKUA
Catharacta maccormicki   to 21"
HABITAT: Breeds in Antarctica; wanders world oceans north to Arctic circle.
NOTES: Large; note pale nape. The more common skua in summer waters off East Coast.

PARASITIC JAEGER
Stercorarius parasiticus   to 18"
HABITAT: Arctic; circumpolar. Wanders to southern tip of S. America in winter.
NOTES: Falconlike. Sharp tail points.

POMARINE JAEGER *Stercorarius pomarinus*
to 22" (twisted tail feathers project 2–7")
HABITAT: Arctic; circumpolar.
NOTES: Largest and bulkiest of 3 jaegers. Often barred below. Distinct projecting twisted tail feathers. White in wing feathers. Light and dark forms.

LONG-TAILED JAEGER
Stercorarius longicaudus
to 23" (thin elongate tail feathers to 10")
HABITAT: Arctic; circumpolar. Winters at sea in S. Hemisphere. Passes well off coast in migration.
NOTES: Most ternlike of jaegers. Small head; adult has very long tail feathers. Little white in upper wings.

PARASITIC
JAEGER
light form

GREAT
SKUA

SOUTH
POLAR
SKUA

PARASITIC
JAEGER
dark form

pale
form

young jaegers,
minus the long
tail-points, are
often distinguish-
able only by size
and build

PARASITIC
JAEGER
immature

PARASITIC
JAEGER
immature

POMARINE
JAEGER
dark form

POMARINE
JAEGER
light form

LONG-TAILED
JAEGER

GULLS Subfamily Larinae

Long-winged swimming birds and master fliers. More robust and with wider wings than terns. Tail is square. **FOOD:** Omnivorous.

GLAUCOUS GULL *Larus hyperboreus* **to 32"**
VOICE: Raucous braying calls.
HABITAT: Arctic; circumpolar. Winter: lakes, coast dumps—any gull congregation site.
NOTES: Huge. In immature birds bill is pink with dark tip. No black in wings; wings end at tail tip. Rare in winter south to dash line.

ICELAND GULL *Larus glaucoides* **26"**
VOICE: Cackling *mews* and *caars*.
HABITAT: Breeds in Arctic. Winter south to Great Lakes and along East Coast.
NOTES: Trimmer than Glaucous. Wingtips are not dark. In immature birds bill is small, blackish. Wingtips extend beyond tail tip. Note gray markings in wingtip feathers. "Kumlien's" form is the race of Iceland Gull that occurs in North America.

IVORY GULL *Pagophila eburnea* **to 17"**
VOICE: High, ternlike *kreeee*.
HABITAT: High Arctic. Circumpolar on ice pack. Wanders south on occasion.
NOTES: Rare visitor south of Gulf of St. Lawrence. Adult: pure white. Young: black-faced and speckled.

ROSS'S GULL *Rhodostethia rosea* **14"**
VOICE: Purrs at nest areas; harsh *caar* when feeding.
HABITAT: Breeds mainly in Siberia and high Arctic.
NOTES: Extremely rare visitor. Breeding adult has a neck-ring and pink chest. Distinct wedge-shaped tail and gray underwings.

"WHITE-WINGED" GULLS
Adults

GLAUCOUS GULL

ICELAND GULL

"Kumlien's" form

typical form

IVORY GULL

ROSS'S GULL

winter

breeding

HERRING GULL *Larus argentatus* to 26"
VOICE: Familar call of the shore: *yuk, yuk, gah, gah.*
HABITAT: Coasts, bays, beaches, lakes, piers, dumps.
NOTES: The classic "seagull," the Herring varies greatly in size. Light gray back, black wingtips. Head is mottled and streaked in winter. Abundant.

THAYER'S GULL *Larus thayeri* to 25"
Very similar to dark Iceland Gulls. Dark eyes. Dark gray, not black, primaries. Rare in East.

RING-BILLED GULL
Larus delawarensis to 19"
VOICE: A high-pitched *kiah yuk, kiyuk kiyuk.*
HABITAT: Coasts, large lakes, estuaries, parks, large suburban parking lots.
NOTES: Adult: distinct ring on bill; yellowish legs.

CALIFORNIA GULL
Larus californicus to 23"
VOICE: A nonstrident *kiak, kiak, kiak.*
HABITAT: Coastal, dumps, fields.
NOTES: Casual to East Coast. Adult: green-yellow legs, dark eyes, red and black spot—not ring—on mandible.

BLACK-LEGGED KITTIWAKE
Rissa tridactyla to 17"
VOICE: A ringing *kaka-week* or *kitti-waak.*
HABITAT: Cliff nester; winters at sea.
NOTES: Small gull; adult's wingtips have a dipped-in-ink appearance.

GREAT BLACK-BACKED GULL
Larus marinus to 31"
VOICE: A deep, harsh *kyow* or *owk.*
HABITAT: Coastal waters, estuaries, large lakes.
NOTES: Massive gull with dark black back.

LESSER BLACK-BACKED GULL
Larus fuscus 23"
Fall and winter visitor appearing more regularly with other gulls along coast and at dumps. Legs are yellow; head is streaked in winter.

THAYER'S GULL

GULLS
Adults

HERRING GULL

HERRING GULL

RING-BILLED GULL

CALIFORNIA GULL

BLACK-LEGGED KITTIWAKE

GREAT BLACK-
BACKED GULL

in winter, gulls on this
plate may be streaked
or clouded with dusky
on crown and hind neck

LESSER BLACK-
BACKED GULL

GREAT BLACK-BACKED GULL

65

LAUGHING GULL *Larus atricilla* **to 17"**

VOICE: A strident laugh: *ha-ha-ha-haah-haah*.
HABITAT: Beaches, salt marshes, bays, piers, ocean.
NOTES: Little white in wingtips. In fall and winter, dusky head marks only. Total hood in spring and summer.

FRANKLIN'S GULL *Larus pipixcan* **to 15"**

VOICE: A shrill *kuk-kuk-kuk*.
HABITAT: Prairies, inland marshes, lakes. Spends winters off South America.
NOTES: Casual on East Coast. White in wingtips separates black from gray.

SABINE'S GULL *Xema sabini* **to 14"**

VOICE: Harsh *carrs* and *peeps* while feeding.
HABITAT: Tundra in summer, ocean in winter.
NOTES: Rare fall visitor. Also occurs on inland lakes migrating across land.

BLACK-HEADED GULL
Larus ridibundus **to 15"**

VOICE: Harsh *carrs* and strident *eeeeks*.
HABITAT: Coastal.
NOTES: Rare but regular on U.S. East Coast and Great Lakes. Recent breeder in Canada. Dark gray underwing.

BONAPARTE'S GULL
Larus philadelphia **to 13"**

VOICE: Nasal, muffled *cherr-cherr*; ternlike peeps.
HABITAT: Oceans, bays, coasts, large inland lakes.
NOTES: Almost ternlike. Rides high in water with wingtips and tail up. Often seen in large groups.

LITTLE GULL *Larus minutus* **to 11"**

VOICE: *Churrs* and nasal chirps.
HABITAT: Lakes and bays.
NOTES: Rare but regular visitor. Look for it in flocks of Bonaparte's Gulls. Tiny but stocky, with black underwings.

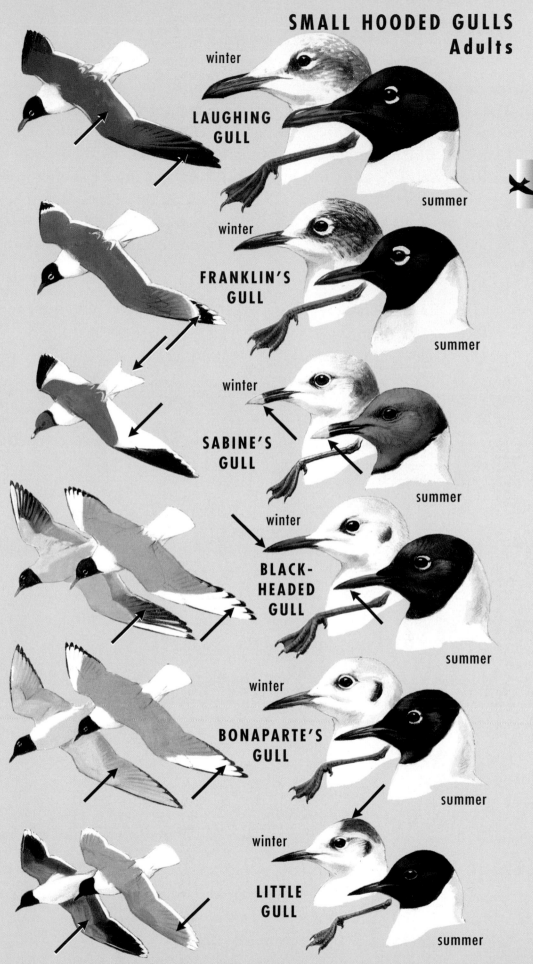

SMALL HOODED GULLS
Adults

winter
LAUGHING
GULL
summer

winter
FRANKLIN'S
GULL
summer

winter
SABINE'S
GULL
summer

winter
BLACK-
HEADED
GULL
summer

winter
BONAPARTE'S
GULL
summer

winter
LITTLE
GULL
summer

67

IMMATURE GULLS (1) LARGER SPECIES

Immature gulls are usually more difficult to identify than adults. They are usually darkest the first year, lighter the second, and in the larger species may not be fully adult until the third or fourth year. Leg and bill colors of most immatures are not as diagnostic as in adults. Identify mainly by pattern and size. The most typical plumages are shown opposite; intermediate stages can be expected. Because of variables such as stage of molt, wear, age, individual variation, and hybridization or albinism, some birds may remain unidentified.

GLAUCOUS GULL   *Larus hyperboreus*   Adult p. 62
FIRST WINTER: Recognized by its larger size, pale tan coloration, and unmarked frosty primaries, a shade lighter than the rest of the wing. Bill is a pale pinkish with a dark tip. **SECOND YEAR:** Very pale or whitish throughout.

ICELAND GULL   *Larus glaucoides*   Adult p. 62
Sequence of plumages similar to Glaucous Gull's, but Iceland Gull is smaller (size of small Herring Gull) with a smaller bill and proportionately longer wings (projecting beyond the tail when at rest). The bill of the first-year Iceland Gull is almost entirely dark.

HERRING GULL   *Larus argentatus*   Adult p. 64
FIRST WINTER: Relatively uniform brown or lightly mottled. Bill shows variable amount of pale at base. **SECOND AND THIRD WINTER:** Paler. Head and underparts whiter; pale gray back; tail feathers dark-tipped, contrast with white rump. Takes four years to mature.

CALIFORNIA GULL   *Larus californicus*   Adult p. 64
FIRST WINTER: Similar to Herring Gull in its first winter but with a shorter bicolored bill.

GREAT BLACK-BACKED GULL   *Larus marinus*   Adult p. 64
Young birds are larger, less brown than first-year Herring Gulls, and have large bills. They are more contrasty ("salt-and-pepper" pattern), have a paler head, tail, and underparts. The "saddle-back" pattern is suggested.

RING-BILLED GULL   *Larus delawarensis*   Adult p. 64
First-winter may be confused with the second-winter Herring Gull, which has a semblance of a ring near the tip of its longer bill, but Ring-billed is smaller, with small bicolored bill.

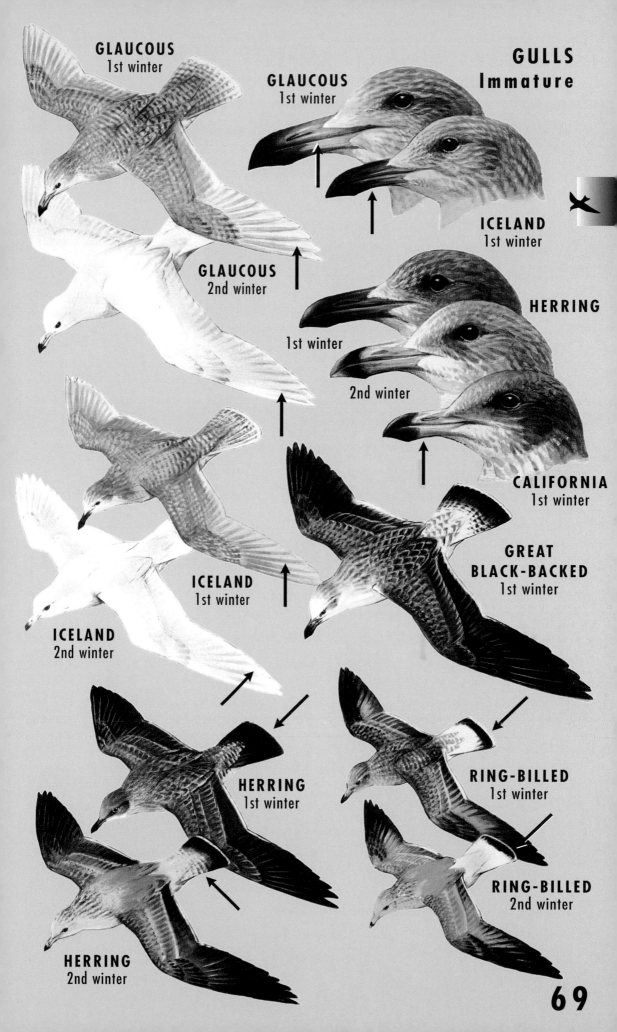

GULLS
Immature

GLAUCOUS
1st winter

GLAUCOUS
1st winter

ICELAND
1st winter

GLAUCOUS
2nd winter

HERRING

1st winter

2nd winter

CALIFORNIA
1st winter

ICELAND
1st winter

GREAT
BLACK-BACKED
1st winter

ICELAND
2nd winter

HERRING
1st winter

RING-BILLED
1st winter

HERRING
2nd winter

RING-BILLED
2nd winter

69

LAUGHING GULL *Larus atricilla* **Adult p. 66**

FIRST YEAR: Dark with a white rump and a white border on the trailing edge of the dark wing. **SECOND WINTER:** Paler or whiter on the chest and forehead; not easy to separate from young Franklin's Gull.

FRANKLIN'S GULL *Larus pipixcan* **Adult p. 66**

The first-winter immature may best be distinguished from the second-winter Laughing Gull by the darker cheek and more distinctly hooded effect and by white outer tail feathers; smaller bill.

BLACK-LEGGED KITTIWAKE *Rissa tridactyla* **Adult p. 64**

Note the dark bar on the nape, black outer primaries, and dark bar across the inner wing. Tail may seem notched.

SABINE'S GULL *Xema sabini* **Adult p. 66**

Dark grayish brown on the back but with the adult's bold triangular wing pattern. Note the lightly forked tail. The young kittiwake is similar but has a dark bar on the nape, a diagonal bar across the wing, and only a slight tail notch.

BONAPARTE'S GULL *Larus philadelphia* **Adult p. 66**

Petite, ternlike. Note the cheek spot, narrow black tail band, and pattern of black and white in the outer primaries.

BLACK-HEADED GULL *Larus ridibundus* **Adult p. 66**

Similar in pattern to immature Bonaparte's Gull, with which it associates, but is slightly larger, less ternlike. Dark undersurface of wing. Reddish or pinkish on bill.

IVORY GULL *Pagophila eburnea* **Adult p. 62**

A ternlike white gull with irregular gray smudges on face, sprinkling of black spots above, and narrow black border on rear edge of its white wings.

LITTLE GULL *Larus minutus* **Adult p. 66**

Smaller than young Bonaparte's with a blacker M pattern formed by the outer primaries and the dark band across the wing. Note especially the dusky cap.

ROSS'S GULL *Rhodostethia rosea* **Adult p. 62**

Note the wedge-shaped, not square or notched, tail and the gray linings of the underwing. Lacks the dark nape of an immature kittiwake. Dovelike head.

1st winter

2nd winter

1st winter

2nd winter

LAUGHING

LAUGHING

1st winter

1st winter

FRANKLIN'S

FRANKLIN'S

SABINE'S

BLACK-LEGGED KITTIWAKE

BONAPARTE'S

BLACK-HEADED

IVORY

LITTLE

ROSS'S

TERNS Subfamily Sterninae

Graceful water birds that are more streamlined than gulls. Bill is sharp, pointed. Tail is usually forked. Terns often hover, then plunge dive to feed or pick food off the surface. **FOOD:** Small fish, marine life.

GULL-BILLED TERN
Sterna nilotica   **to 14"**

VOICE: A throaty, raspy *za-za-za*; also *kay-weck*.
HABITAT: Salt marshes, fields, coastal bays, freshwater ponds.
NOTES: Heavy, almost gull-like bill. Shallow notch in tail. Hunts insects over fields and marshes. Very white in flight.

SANDWICH TERN
Sterna sandvicensis   **to 18"**

VOICE: A grating, high-pitched *kirr-ick*.
HABITAT: Coastal waters, jetties, beaches.
NOTES: Breeding birds have distinct crest. The Sandwich is the only tern with yellow-tipped black bill. Breeding colonies are very local.

ROYAL TERN   *Sterna maxima*   **to 21"**

VOICE: A distinct *cheer-eek*; also *kaah*.
HABITAT: Coasts, sandy beaches, salt bays.
NOTES: Large, trim tern. Light orange bill. Crest. White forehead except in breeding season (April–June). Roams northward after breeding.

CASPIAN TERN   *Sterna caspia*   **to 23"**

VOICE: A rasping, low *kraa-uh* or *karr*. Also *kak, kak*. Immature call a higher-pitched whistle.
HABITAT: Large lakes, coastal waters, beaches, bays.
NOTES: Large stocky tern with a red-orange bill. Modest crest. Forehead speckled in winter. Rranges inland to lakes whereas Royal does not.

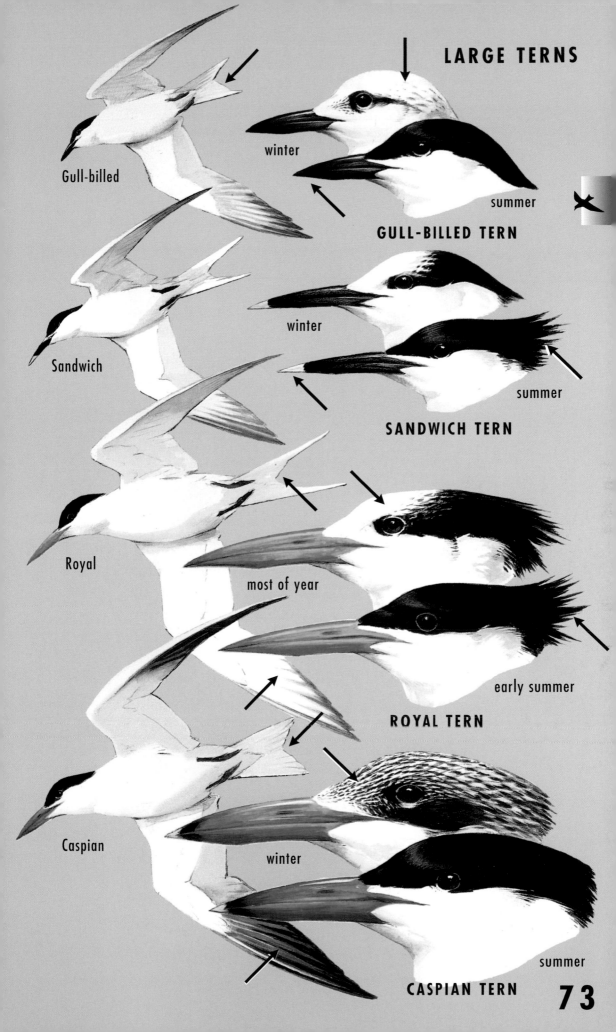

LARGE TERNS

Gull-billed

winter

summer

GULL-BILLED TERN

Sandwich

winter

summer

SANDWICH TERN

Royal

most of year

early summer

ROYAL TERN

Caspian

winter

summer

CASPIAN TERN

73

LEAST TERN *Sterna antillarum* 9"

VOICE: A sharp, repeated *kit* or *kitti-kitti-kitti*.
HABITAT: Sea beaches, bays, large rivers, sandbars.
NOTES: Small; yellow bill and white forehead. The Least Tern is an endangered species in many regions because of loss of nesting areas and human disturbance.

ARCTIC TERN *Sterna paradisaea* to 17"

VOICE: *Kee-yah;* also characteristic *keer-keer*.
HABITAT: Open ocean and tundra ponds.
NOTES: Travels from Arctic to Antarctic every year. Blood red bill. Grayish body contrasts to white cheek and black cap. Short legs. Rarely seen south of Massachusetts.

COMMON TERN *Sterna hirundo* to 16"

VOICE: A harsh, downward *kee-arr kik-kik-kik*.
HABITAT: Oceans, bays, beaches, lakes. Nests colonially on sandy beaches and marshes.
NOTES: Slim; black cap, grayish underparts, orange-red bill with black tip.

FORSTER'S TERN *Sterna forsteri* to 15"

VOICE: A soft, nasal *zzuur* or nasal *kik . . . kik . . . kik*.
HABITAT: Marshes (nests), lakes, bays, beaches.
NOTES: Frosty wingtips. In fall and winter loses full cap and has oval-shaped black mark through eye.

ROSEATE TERN *Sterna dougallii* to 17"

VOICE: Rasping *ka-a-ak*, distinct *chuk-ittuk*.
HABITAT: Coastal; salt bays, estuaries, ocean.
NOTES: When at rest, its long white tail feathers extend well beyond wingtips. Pink bloom to adult breast. Thin black bill. Wingbeats shallow and rapid. Loss of breeding sites has made this a federally endangered species.

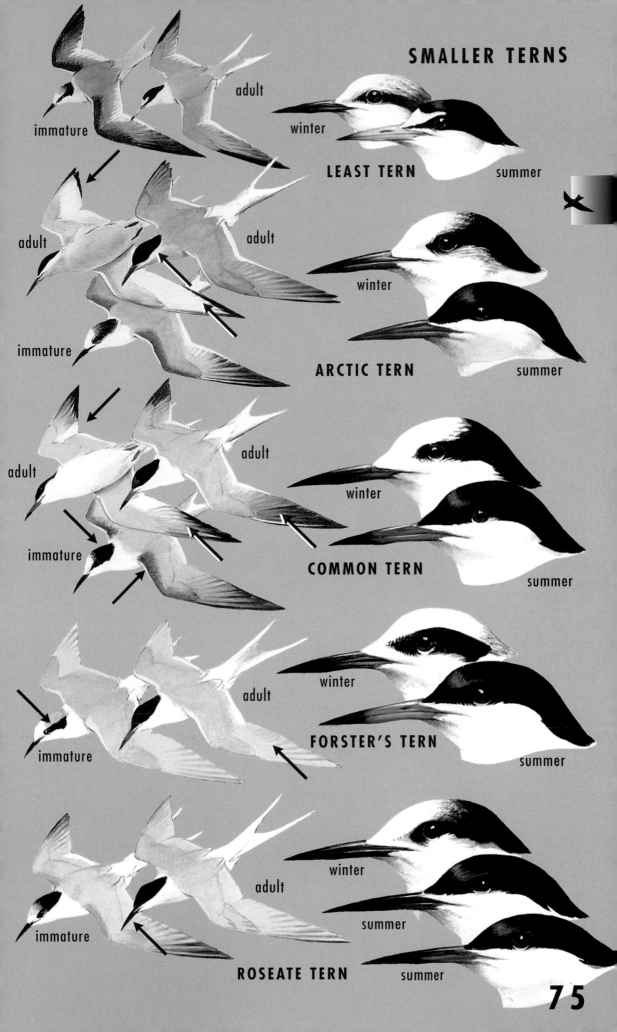

SMALLER TERNS

immature

adult

winter

LEAST TERN

summer

adult

adult

immature

winter

ARCTIC TERN

summer

adult

adult

immature

winter

COMMON TERN

summer

immature

adult

winter

FORSTER'S TERN

summer

immature

adult

winter

summer

ROSEATE TERN

summer

BLACK TERN *Chlidonias niger* to 10"
VOICE: A sharp *kik, keek,* or *klea.*
HABITAT: Freshwater lakes and marshes. Coastal waters in migration.
NOTES: Our only black-bodied tern. Smudged appearance when molting. Pied markings in winter.

BROWN NODDY *Anous stolidus* to 15"
VOICE: A ripping *karrrrk* or *arrrrowk.*
HABITAT: Tropical oceans.
NOTES: Brown tern with wedge-shaped tail. Distinct white cap and dark body. Breeds on Dry Tortugas. Carried north by hurricanes. Black Noddy (*A. tenuirostris*) can occur with Brown on Dry Tortugas.

SOOTY TERN *Sterna fuscata* to 17"
VOICE: A nasal *wide-a-wake* or *wacky-wack.*
HABITAT: Pantropical oceans.
NOTES: Breeds on Dry Tortugas; often carried north by hurricanes. Has been found well inland after these storms. Also breeds in Louisiana and North Carolina.

BRIDLED TERN *Sterna anaethetus* to 14"
VOICE: A sharp *kyee kit kit.*
HABITAT: Tropical oceans.
NOTES: Gray- (not black-) backed with white collar around neck and whitish tail. Regular in Gulf Stream north to at least North Carolina; hurricanes may carry some farther north or inland.

SKIMMERS Family Rynchopidae

Has a knifelike red flattened bill and uses longer lower mandible to skim food from surface of water. **FOOD:** Small fish and crustaceans.

BLACK SKIMMER *Rynchops niger* to 20"
VOICE: Short barking notes: *yaap, yaap, yip.*
HABITAT: Ocean beaches, salt bays, tidewaters.
NOTES: Large groups yap like small dogs.

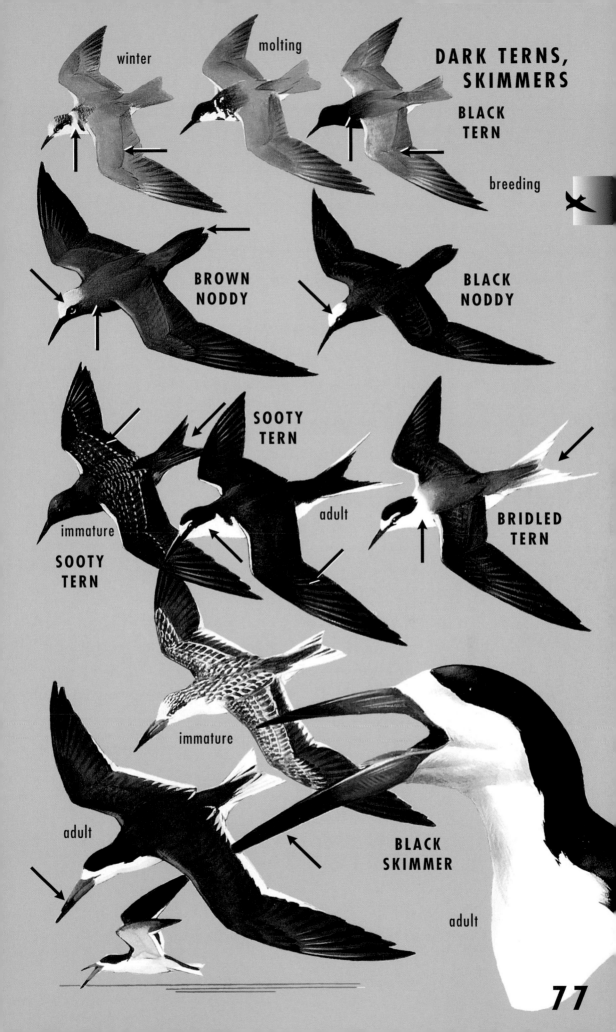

DARK TERNS,
SKIMMERS

BLACK
TERN

winter

molting

breeding

BROWN
NODDY

BLACK
NODDY

SOOTY
TERN

immature

adult

SOOTY
TERN

BRIDLED
TERN

immature

adult

BLACK
SKIMMER

adult

77

HERONS, BITTERNS Family Ardeidae

Medium to large wading birds with long necks and spearlike bills. Stand with head erect or tucked on shoulders. In flight, legs trail and neck folds in S position. FOOD: Fish, frogs, crayfish, other aquatic life, mice, snakes, insects.

GREAT BLUE HERON *Ardea herodias* to 52"
VOICE: Deep, harsh croaks: *frahnk, frahnk*.
HABITAT: Marshes, swamps, shores, tide flats.
NOTES: Nests locally in colonies of large stick nests.

"WURDEMANN'S HERON" *Ardea herodias*
This color morph of the Great Blue is a bird of the Florida Keys. Note all-white head.

LITTLE BLUE HERON *Egretta caerulea* to 24"
VOICE: A loud, harsh croaking with guttural grunts.
HABITAT: Marshes, swamps, rice fields, ponds, shores.
NOTES: Goes through 3 plumage changes: pure white to piebald to adult (see page 81). Wanders north in summer to dash line.

TRICOLORED HERON
Egretta tricolor   26"
VOICE: Harsh, croaking *cut cut* notes, squawks.
HABITAT: Marshes, swamps, ponds, shores.
NOTES: Its dark neck contrasts with its white belly. Long head plumes appear in breeding season. Occasionally nests north of mapped range. Wanders north to dash line.

REDDISH EGRET
Egretta rufescens   29"
VOICE: Croaks, squawks, growls.
HABITAT: Coastal tide flats, salt marshes, lagoons.
NOTES: When searching for food, the Reddish Egret often jumps about in staggering and lurching fashion with wings outstretched. Has an all-white color morph (see page 81).

Herons fly with necks pulled in

Herons' necks may stretch or be looped in

immature

adult

GREAT BLUE HERON

adult

"WURDEMANN'S" HERON

adult

LITTLE BLUE HERON
(White immature on p. 81)

adult

TRICOLORED HERON

REDDISH EGRET

(white color morph on p. 81)

Reddish Egret "dancing"

79

GREAT EGRET *Ardea alba*   38"
VOICE: Low, hoarse croak; also rapid *cuk, cuk, cuk.*
HABITAT: Marshes, ponds, shores, mud flats.
NOTES: Yellow bill, black legs. Lime green of facial skin intensifies in spring. Stalks food. Post-breeding dispersal to dash line.

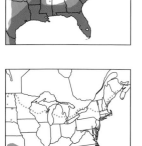

SNOWY EGRET *Egretta thula*   to 27"
VOICE: Low croaking; in colony, bubbly *wulla, wulla.*
HABITAT: Marshes, swamps, ponds, shores, tide flats.
NOTES: Thin black bill and "golden slipper" feet. Beautiful back plumes in breeding season. Occasional nesting north to dash line. Post-breeders wander widely.

LITTLE BLUE HERON
Shown here with white egrets. Note change from pure to piebald (see p. 78).

CATTLE EGRET *Bubulcus ibis*   20"
VOICE: Raucous squawks, guttural *kacks* at colony.
HABITAT: Farms, marshes, highway edges. Associates with cattle or follows ground tillers.
NOTES: Immigrated to S. America from Africa, then to N. America. Reddish blush to head, chest, and back in breeding condition. Stands with hunched shoulders. Nests and wanders locally and sporadically north to dash line.

REDDISH EGRET
White morph. Note 2-toned bill in breeding season. (See p. 78.)

"GREAT WHITE" HERON
Ardea herodias   to 50"
VOICE: Guttural croaking.
HABITAT: Mangrove keys, salt bays, open mud flats, marshes.
NOTES: White morph of the Great Blue. In U.S., found only in Florida. Note yellow bill and legs. Largest white heron in N. America.

WHITE
HERONS,
EGRETS

GREAT
EGRET

SNOWY
EGRET

changing

immature

LITTLE BLUE
HERON
(adult on p. 79)

CATTLE
EGRET
immature

breeding

REDDISH
EGRET

white
morph

"GREAT
WHITE"
HERON
(s. Florida
color morph of
Great Blue
Heron)

(dark morph
on p. 79)

81

BLACK-CROWNED NIGHT-HERON
Nycticorx nycticorax   **to 28"**
VOICE: Deep, throaty *quok* or *quark*.
HABITAT: Roosts in trees. Marshes, shores, breakwaters.
NOTES: Adult has black cap. Immature: brown with large spots; eyes yellowish. In flight, feet extend just beyond tail. Often hunts at inland ponds and lakes by night. Colonies very local.

YELLOW-CROWNED NIGHT-HERON
Nyctanassa violacea   **to 28"**
VOICE: Has higher-pitched *quark* than Black-crowned.
HABITAT: Cypress swamps, mangroves, bayous, marshes, streams, ponds, lakes.
NOTES: Adult has gray back, yellowish crown. In flight, feet extend well beyond tail end. Immature grayer than Black-crowned, thicker, all-dark bill, and orange-red eye. More common in South.

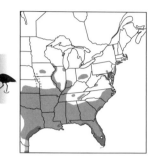

GREEN HERON *Butorides virescens*   **to 22"**
VOICE: A loud *skyow* and a series of *kuks*.
HABITAT: Lakes, ponds, marshes, streams, swamps.
NOTES: Our most widespread heron inland. Often keeps chestnut neck tucked in; occurs singly or in pairs.

LEAST BITTERN *Ixobrychus exilis*   **to 14"**
VOICE: Low, muted *coo-coo-coo*, sharp *kack, kack*.
HABITAT: Fresh marshes, reedy pond edges, cattail marshes. Difficult to flush.
NOTES: Very small. Straddles reeds as it moves about; in flight shows distinct buffy wing patches. Very local within mapped range.

AMERICAN BITTERN
Botaurus lentiginosus   **23"**
VOICE: A booming, slow *oong-ka-choonk, oong-ka-choonk*. When flushed: *kok-kok-kok*.
HABITAT: Marshes, reedy lakes, wet field edges, prairies, salt marshes in winter.
NOTES: Will freeze in position rather than fly. Rare and irregular breeder south of mapped range.

HERONS, BITTERNS

adult

BLACK-CROWNED NIGHT-HERON

immature

adult

YELLOW-CROWNED NIGHT-HERON

immature

typical

dark morph (rare)

LEAST BITTERN

immature

adult

GREEN HERON

AMERICAN BITTERN

83

STORKS Family Ciconiidae

Large, long-legged, heronlike birds with long bills. Some have naked heads. Gait is a sedate walk. **FOOD:** Frogs, crustaceans, lizards, rodents.

WOOD STORK *Mycteria americana*
to 47" (wingspan to 5½')

VOICE: Hoarse croak, especially at nesting colonies.
HABITAT: Nesting colonies in cypress swamps; marshes, ponds, lagoons.
NOTES: Our only native stork, the Wood Stork has a distinct naked head. It often rides thermals high into sky. Flies with neck outstretched. Post-breeders wander north to dash line.

CRANES Family Gruidae

Stately birds; more robust than herons. Cranes often have red facial skin. Note tufted feathers over rump. Cranes fly with neck extended and distinct upward snap of the wing. **FOOD:** Omnivorous.

WHOOPING CRANE *Grus americana* to 50"

VOICE: A trumpeting *ker-loo! keer-lee-oo!*
HABITAT: Summer: muskeg; winter: coastal marshes. In migration may stop in grain fields.
NOTES: Tallest American bird. Adult is white with red face; immature has rusty head. One of the rarest birds in the U.S.—extensive breeding and reintroduction programs are under way. Winter location is Aransas Refuge, Texas. Breeds in Northwest Territories.

SANDHILL CRANE *Grus canadensis*
to 48" (wingspan to 7')

VOICE: A bugling *garooooo-a-a-a.*
HABITAT: Summer: tundra; prairies, fields, marshes, river bottoms. Winter: grain fields, prairie, marshes.
NOTES: Congregate in huge numbers in prairie river bottoms in Nebraska during migration north.

below, for comparison:
left, White Ibis
right, Wood Stork

adult

WOOD
STORK

immature

Storks, ibises, and
cranes fly with
necks outstretched

adult

immature

WHOOPING
CRANE

adult

immature

SANDHILL
CRANE

LIMPKINS Family Aramidae

One species in the family through N., Cen., and S. America. FOOD: Freshwater snails (genus *Pomacea*), few insects, frogs.

LIMPKIN *Aramus guarauna*   **28"**
VOICE: Piercing, repeated *kree-ow, kree-ow*. Frequently calls at night and on overcast days.
HABITAT: Cypress swamps, freshwater marshes.
NOTES: A large spotted swamp wader. Flight is cranelike with an upward wing snap.

IBISES and SPOONBILLS Family Threskiornithidae

Long-legged, medium-sized, heronlike birds with down-curved bills. Spoonbills have spatulate bills. FOOD: Small crustaceans, fish, insects.

GLOSSY IBIS *Plegadis falcinellus*   **to 25"**
VOICE: A guttural *ka-onk*.
HABITAT: Marshes, rice fields, swamps.
NOTES: Adult: Deep chestnut color with glossy bronzy hues. Flies in lines with several quick flaps, then glides. Thin light blue border to face. A few wander inland to dash line.

WHITE-FACED IBIS *Plegadis chihi*   **to 25"**
VOICE: A guttural growl.
HABITAT: Marshes, flooded fields, swamps.
NOTES: White stripe around base of facial skin (adult), red eyes, reddish legs in spring. It is extending its range eastward, regular to dash line.

WHITE IBIS *Eudocimus albus*   **to 27"**
VOICE: Grunts, croaks, and squabbling growls.
HABITAT: Salt, brackish, and freshwater marshes; mangrove swamps. Often seen in large groups.
NOTES: Red face and long, curved red bill. Young splotchy brown and white. Flies in lines. Disperses in summer north to dash line.

LIMPKIN,
IBISES

below, for facial
comparison in
breeding season

WHITE-
FACED
IBIS

GLOSSY
IBIS

LIMPKIN

GLOSSY
IBIS
adult

GLOSSY
IBIS
immature

WHITE IBIS
adult

WHITE IBIS
immature

87

SCARLET IBIS *Eudocimus ruber*   25"

This ibis is an unmistakable scarlet color. Possible accidental to Gulf states. Most birds are escapes from zoos and are pink, not scarlet, in color (they often lose color in captivity).

ROSEATE SPOONBILL *Ajaia ajaja*   32"

VOICE: Low grunting croak at nesting colony.

HABITAT: Coastal marshes, lagoons, mangrove swamps, and islands (where it nests).

NOTES: Flat, spoonlike bill and brilliant pink color make adults unmistakable. Immature birds are white but with spoon-shaped bill. Sweeps bill back and forth while feeding. Disperses to dash line in late summer.

FLAMINGOS Family Phoenicopteridae

One of the best-known birds. Pink, with a long, curved neck, long legs, and a heavy, sharply bent bill. **FOOD:** Crustaceans, blue-green algae, diatoms, and some mollusks.

GREATER FLAMINGO

Phoenicopterus ruber   45"

VOICE: A gooselike trumpet: *ar-honk*; rolling purrs while feeding.

HABITAT: Salt flats, saline lagoons.

NOTES: Most birds reported are pale, washed-out zoo escapes. Small population in Florida Bay. May be wild visitors from the Bahamas.

SCARLET IBIS

adult

ROSEATE SPOONBILL

immature

adult

adult

GREATER FLAMINGO

RAILS, GALLINULES, MOORHENS, and COOTS
Family Rallidae (in part)

Rails are compact, hen-shaped marsh birds. Usually secretive and heard more often than seen. Flight is brief and reluctant, with legs dangling. (The coot, moorhen, and gallinule have been treated separately on page 42, but heads are described below.) **FOOD:** Aquatic plants, seeds, insects, small frogs, crustaceans, mollusks.

VIRGINIA RAIL *Rallus limicola*   **9"**
VOICE: *Wak, awak, awak,* in descending pattern. Also *kidick, kidick* and various grunting sounds.
HABITAT: Fresh and brackish marshes; salt marshes during the winter.
NOTES: A small rusty rail that looks blackish to quick view. Runs with neck outstretched. Rare breeder south of range map.

KING RAIL *Rallus elegans*   **to 19"**
VOICE: Slow, deep, spaced *rit, rit, rit, rit* or clacking *yit, yit, yit,* not descending.
HABITAT: Fresh and brackish marshes, flooded fields, rice fields, swamps. Salt marshes in winter.
NOTES: Large. Rich rufous, including cheek; rufous "shoulder," back boldly patterned. Very local breeder in much of range, sporadically north to dash line.

CLAPPER RAIL *Rallus longirostris*   **to 16"**
VOICE: A clattering, echoing, clapping *kek-kek-kek-kek,* often accelerating in pace and then trailing off.
HABITAT: Salt (sometimes brackish) marshes, and locally in mangroves.
NOTES: A common marshland bird that is heard far more often than seen. Overall grayish brown with distinct gray cheeks; Gulf Coast birds richer colored, more like King Rail.

LONG-BILLED RAILS

adult

immature

VIRGINIA
RAIL

KING
RAIL

chick

CLAPPER RAIL

Limpkin
for comparison

91

SORA *Porzana carolina* — to 9¾"

VOICE: A descending whinny; in spring a plaintive *ker-whee?* Also a loud *keek* when alarmed.

HABITAT: Freshwater marshes, wet meadows, grain fields, salt marshes in winter.

NOTES: A small rail with stubby bill and bold, striped sides. A rare breeder south of mapped range.

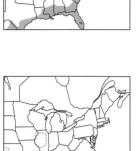

BLACK RAIL *Laterallus jamaicensis* — to 6"

VOICE: Distinct *kiki-doo* or *dee-dee-doo*.

HABITAT: Tidal and freshwater marshes, glasswort (*Salicornia*) flats, grassy marshes.

NOTES: Beware: All rails have black young, which are often misidentified as Black Rails. Black Rails, sparrow-sized skulkers with barred flanks, run about, mouselike. Usually only a voice in the marsh. Status in interior is not well known.

YELLOW RAIL
Coturnicops noveboracensis — to 7"

VOICE: Nocturnal ticking notes that sound like 2 nickels being struck together: *tic-tic tic-tic-tic* in groups of 2–3.

HABITAT: Grassy freshwater marshes, sedge meadows, grain fields. In winter wild rice fields and upper salt marshes.

NOTES: Small and buffy and blackish, like a week-old chick. Mouselike and difficult to flush. Bold white wing patch in flight. Breeds very locally within dash line.

CORN CRAKE *Crex crex* — to 10"

A visitor from Europe through very early 1900s. Only a handful of records since then. Population declining in Europe.

COOT, MOORHEN, AND GALLINULE: Separation by frontal shield coloration. See p. 42 for complete descriptions.

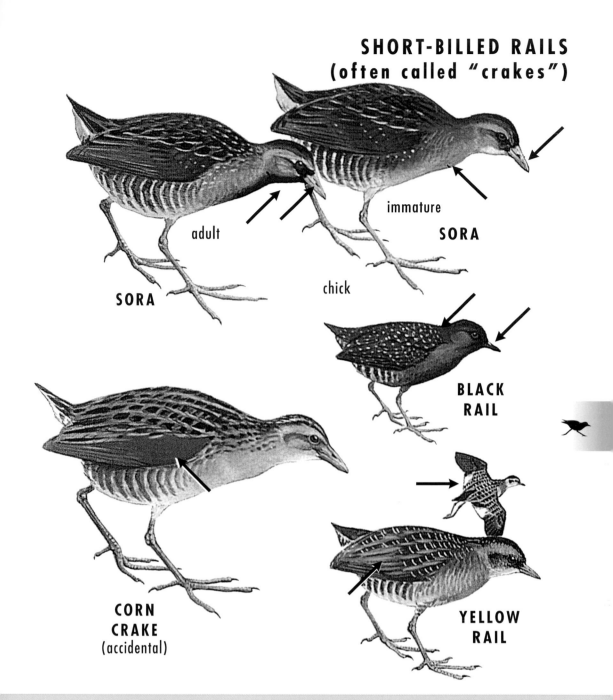

SHORT-BILLED RAILS
(often called "crakes")

SORA

adult

chick

SORA

immature

SORA

BLACK
RAIL

CORN
CRAKE
(accidental)

YELLOW
RAIL

COOT, MOORHEN, GALLINULE

AMERICAN
COOT

COMMON
MOORHEN

PURPLE
GALLINULE

(swimmers related to rails; see p. 43)

OYSTERCATCHERS Family Haematopodidae

Large waders with long, laterally flattened, chisel-tipped red bills. **FOOD:** Mollusks, crabs, marine worms.

AMERICAN OYSTERCATCHER
Haematopus palliatus   **to 21"**
VOICE: A piercing *wheep* or *kleep*; also *pic-pic*.
HABITAT: Coastal beaches, tidal flats.
NOTES: Distinctive. Bold black and white wing pattern in flight. Strictly a coastal species.

AVOCETS and STILTS Family Recurvirostridae

Slim waders with long thin legs and slender bills. Avocets have a distinct upcurved bill. **FOOD:** Insects, crustaceans, other aquatic life.

BLACK-NECKED STILT
Himantopus mexicanus

to 17" (including legs)
VOICE: A sharp yipping: *kyip-kyip-kyip*.
HABITAT: Grassy marshes, mud flats, pools, lake edges.
NOTES: Trim black and white bird with exceptionally long pink legs. Very reactive when people are nearby: continuously calling, scolding and diving.

AMERICAN AVOCET
Recurvirostra americana   **to 20"**
VOICE: A sharp *wheek* or *kleet,* excitedly repeated.
HABITAT: Mud and alkaline flats, shallow lakes, prairie ponds.
NOTES: Upturned bill is unique. Bold black and white pattern set off by orangey or russet-tan head (grayish in winter). Feeds by swinging bill back and forth like a scythe. Often feeds in large groups.

OYSTERCATCHER,
STILT, AVOCET

AMERICAN
OYSTERCATCHER

BLACK-NECKED
STILT

summer

winter

AMERICAN AVOCET

95

PLOVERS Family Charadriidae

Shorebirds, plovers are more compactly built, thicker-necked, and shorter-billed than most sandpipers. Plovers feed like robins: walk, stop, pick food on mud flats and sandbars. Ruddy Turnstone now assigned to sandpipers. **FOOD:** Small marine life, some insect matter.

BLACK-BELLIED PLOVER
Pluvialis squatarola   **to 13"**

VOICE: Distinctive, plaintive whistle: *pee-o-wee* or *tlee-oo-eee* (middle note lower).

HABITAT: Mud flats, marshes, plowed fields, beaches; nests in open tundra.

NOTES: Distinct in summer plumage but may appear in winter plumage at any time of year. Note stout bill. White rump and black underwings show in flight.

AMERICAN GOLDEN-PLOVER
Pluvialis dominica   **to 11"**

VOICE: Whistled *queedle* or *que-e-a* dropping at end.

HABITAT: Prairies, short grass, mud flats, shores; nests on tundra.

NOTES: Adults are golden-backed with black extending under the tail. Distinguish winter-plumage birds from Black-bellied Plover by smaller bill, capped effect to top of head, plain rump, and gray underwings. Spring migration mostly through interior, fall migration also on Atlantic Coast.

RUDDY TURNSTONE
Arenaria interpres   **to 10"**

VOICE: A staccato *chuck-a-tuck* or a single *kewk*.

HABITAT: Beaches, mud flats, jetties, rocky shores; nests in tundra.

NOTES: Bold harlequin pattern in breeding plumage. Retains yellow-orange legs in winter and has smudged chest. Bold white wing pattern is distinct. Uses upturned bill to forage for horseshoe crab eggs and flip over stones. Scarce inland.

winter

winter

BLACK-
BELLIED
PLOVER

BLACK-
BELLIED
PLOVER

summer

winter

winter

winter

AMERICAN
GOLDEN-
PLOVER

AMERICAN
GOLDEN-
PLOVER

summer

winter

summer

RUDDY
TURNSTONE

97

SEMIPALMATED PLOVER

Charadrius semipalmatus 7½"

VOICE: Plaintive, upward-slurred *chi-wee* or *to-lit*.
HABITAT: Shore, tidal flats. Nests in tundra.
NOTES: Has a complete neck-ring and a brown back. Single breast band. Partial toe-webbing.

COMMON RINGED PLOVER

Charadrius hiaticula 7½"

Accidental visitor from Europe. Almost identical to Semipalmated, but toes are not webbed and breast band is wider. Best told by call: a soft whistled tooip.

PIPING PLOVER

Charadrius melodus 7½"

VOICE: A plaintive whistle: *peep-peep-peep* and a 2-note *peep-lo*.
HABITAT: Sand beaches and tidal flats.
NOTES: Sand-colored, with incomplete neck-ring. Federally endangered. Breeds very locally within dash line.

SNOWY PLOVER

Charadrius alexandrinus 6½"

VOICE: A musical *pe-wee-ah* or *o-wee-ah*.
HABITAT: Beaches and sandy flats.
NOTES: Pale plover of the Gulf Coast. Slim black (*not* 2-toned) bill, darkish legs.

WILSON'S PLOVER

Charadrius wilsonia to 8"

VOICE: An emphatic whistled *whit!* or *wheet!*
HABITAT: Open beaches, tidal flats, sandy islands.
NOTES: Heavy black bill, wide chest band, gray-colored legs. Wanders north with some regularity.

KILLDEER *Charadrius vociferus* to 11"

VOICE: Loud, insistent *kill-deeah* and a plaintive *dee-dee-dee* (rising) or *trill* near nest.
HABITAT: Fields, airports, lawns, sandy parking areas, plowed fields, riverbanks, pond edges, marshes.
NOTES: The only 2-ringed plover. Orange rump displayed in flight or while giving broken-wing act at nest to lead away predators.

BELTED
PLOVERS

COMMON
RINGED
PLOVER

SEMIPALMATED
PLOVER

winter

summer

winter

elted form
PIPING
PLOVER

summer

immature ♀

♂

summer

SNOWY
PLOVER

♀ ♂

WILSON'S
PLOVER

KILLDEER

chick

HEADS OF BELTED PLOVERS

Piping

Ringed

Wilson's

Snowy

Semipalmated

Killdeer

99

PLOVERS AND TURNSTONE IN FLIGHT

Learn the distinctive flight-calls.

PIPING PLOVER *Charadrius melodus*   **p. 98**
Note is a plaintive, whistled *peep-lo* (first note is higher).

SNOWY PLOVER *Charadrius alexandrinus*   **p. 98**
A pale sand color. Tail has a dark center, sides are white, rump is not white. Note is a whistle: *pe-wee-ah* or *o-wee-ah*.

SEMIPALMATED PLOVER *Charadrius semipalmatus*   **p. 98**
Mud brown; dark tail with white borders. Note is a plaintive, upward-slurred *chi-we* or *too-li*.

WILSON'S PLOVER *Charadrius wilsonia*   **p. 98**
Larger than the Semipalmated yet has a similar pattern and a large bill. Note is an emphatic, whistled *whit!* or *wheet!*

KILLDEER *Charadrius vociferus*   **p. 98**
Tawny-orange rump, longish tail. Noisy; note is a loud *kill-deeah* or *killdeer; dee-dee-dee*, etc.

BLACK-BELLIED PLOVER *Pluvialis squatarola*   **p. 96**
Spring: Black below, white undertail coverts. Pattern above as in fall. Fall: Black axillars ("wingpits"), white in wing and tail. Note is a plaintive, slurred whistle, *whee-er-eee*.

AMERICAN GOLDEN-PLOVER *Pluvialis dominica*   **p. 96**
Spring: Black below, black undertail coverts. Fall: Underwing grayer than Black-bellied's; no black in axillars. Note is a harsh, whistled *quee-dle* or *quee*.

RUDDY TURNSTONE *Arenaria interpres*   **p. 96**
Harlequin pattern. Note is a low, chuckling *ket-a-kek* or *kut-a-kut*.

PIPING
PLOVER

SNOWY
PLOVER

KILLDEER

SEMIPALMATED
PLOVER

WILSON'S
PLOVER

winter

winter

BLACK-
BELLIED
PLOVER

summer

BLACK-
BELLIED
PLOVER

winter

winter

AMERICAN
GOLDEN-
PLOVER

summer

AMERICAN
GOLDEN-
PLOVER

summer

RUDDY
TURNSTONE

101

SANDPIPERS Family Scolopacidae

Small to medium-sized waders. Bills are longer and more slender than plovers'. FOOD: Insects, crustaceans, mollusks, worms, etc.

AMERICAN WOODCOCK
Scolopax minor   **to 11"**
VOICE: Dusk song in spring. Sounds a chirping note while spiraling upward; wing twitters as it descends. On ground a nasal *beeezp* or *beeent*.
HABITAT: Wet thickets, moist woods, brushy swamps.
NOTES: Displays over open fields. Rotund; long bill.

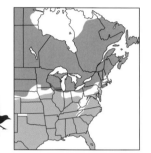

COMMON SNIPE *Gallinago gallinago*   **11"**
VOICE: A rasping *scaip*. A "winnowing" is created by its tail as it circles high overhead.
HABITAT: Marshes, bogs, wet meadows.
NOTES: Very long bill. Relies on camouflage as it sits tight in grasses. Erratic zigzag flight.

SHORT-BILLED DOWITCHER
Limnodromus griseus   **to 12"**
VOICE: A staccato *tu-tu-tu*.
HABITAT: Mud flats, tidal marshes, pond edges.
NOTES: Best distinguished from Long-billed by call.

LONG-BILLED DOWITCHER
Limnodromus scolopaceus   **to 12½"**
VOICE: A single *keek* (sometimes in a series).
HABITAT: Prefers freshwater lakes and pond margins in migration. Mud flats in winter.
NOTES: Regular migrant through interior; travels to Atlantic Coast in fall. Winters on Gulf Coast and north to New Jersey.

RED KNOT *Calidris canutus*   **to 11"**
VOICE: A low *knut* and a mellow *tooit-wit*.
HABITAT: Tidal flats, sandy shores. Tundra in summer.
NOTES: Stocky with short bill. Looks like a large Sanderling. Visits specific sites (e.g., coast of Delaware Bay) for mass buildups during migration.

SNIPE,
WADERS,
etc.

AMERICAN
WOODCOCK

COMMON
SNIPE

winter

SHORT-
BILLED
DOWITCHER

summer

winter

summer

LONG-
BILLED
DOWITCHER

winter

summer

RED KNOT

winter

HUDSONIAN GODWIT
Limosa haemastica   **to 16"**
VOICE: A sharp, high *ta-wit* or *godwit!*
HABITAT: Beaches, mud flats, prairies, rice fields.
NOTES: Prairie migration north in spring. Fall migration on Northeast coast well offshore.

MARBLED GODWIT *Limosa fedoa*   **to 20"**
VOICE: Accented *kerwhit* or *god-wit*, also *raddica*.
HABITAT: Prairies, pools, shores, tidal flats, beaches.
NOTES: A prairie breeder that is rare on the East Coast and Great Lakes. Transient through interior.

LONG-BILLED CURLEW
Numenius americanus
to 26" (including 4–8½" downcurved bill)
VOICE: A loud *cur-lee* or whistled *kil-li-li.*
HABITAT: Plains and rangeland. Winter: beaches, tidal flats, salt marshes, and cultivated land.
NOTES: Rare in winter on Gulf and southeastern coast and accidental to northeastern coast where at one time it was a transient. Loss of prairie has pushed it farther west.

ESKIMO CURLEW *Numenius borealis* **to 14"**
This small curlew is believed to be extinct. Bred in Arctic and traveled north through Great Plains in spring, south along East Coast in fall. May easily be confused with Whimbrel. Small, thin bill and cinnamon underwings diagnostic.

WHIMBREL *Numenius phaeopus*   **to 19"**
VOICE: 5–7 short rapid whistles: *ti-ti-ti-ti-ti-ti.*
HABITAT: Shores, beaches, mud flats, salt marshes, tundra.
NOTES: Groups fly in lines often well offshore when migrating south. European race has white rump. Rare inland.

LARGE
SANDPIPERS

summer

winter

HUDSONIAN
GODWIT

winter

♂ summer

MARBLED
GODWIT

LONG-
BILLED
CURLEW

ESKIMO
CURLEW
(underwing)

WHIMBREL

105

WILLET
Catoptrophorus semipalmatus   **to 17"**
VOICE: A musical, repeated *pill, will, willet* on breeding grounds.
HABITAT: Salt marshes, beaches, wet meadows, mud flats, fresh marshes.
NOTES: Heavy bill, dark legs, and bold wing pattern separate it from yellowlegs. Very protective at nest site and will dive-bomb intruders.

GREATER YELLOWLEGS
Tringa melanoleuca   **14"**
VOICE: A 3-note whistle: *twhew-twhew-twhew.*
HABITAT: Open marshes, mud flats, streams and pond edges.
NOTES: Has a distinctive voice, white rump, long yellow legs. Bill curves slightly upward.

LESSER YELLOWLEGS *Tringa flavipes* **to 11"**
VOICE: 1- to 2-note *yew* or *yu-yu.*
HABITAT: Marshes, mud flats, shores, pond edges. Breeding grounds in open boreal woods.
NOTES: Smaller than Greater Yellowlegs and has a shorter, slimmer, straighter all-dark bill. Voice is the best distinction.

SOLITARY SANDPIPER
Tringa solitaria   **to 9"**
VOICE: A sweet, sharp *peet* or *peet-weet-weet.*
HABITAT: Streamsides, ponds, open marshes.
NOTES: Very dark with sharply contrasting white outer tail feathers. Flies with quick, stiff wingbeats followed by glides.

NOTE: Winter-plumage Stilt Sandpiper (see page 110) and Wilson's Phalarope (see page 114) shown here, as they have a similar appearance to shorebirds on plate. Stilt has a white rump and distinct eyebrow. Bill curves downward slightly at tip. Wilson's Phalarope has a needlelike bill. Darts quickly about for food.

SANDPIPERS

summer

WILLET

winter

LESSER
YELLOWLEGS

GREATER
YELLOWLEGS

(in flight
on p. 119)

SOLITARY
SANDPIPER

STILT
SANDPIPER

winter

WILSON'S
PHALAROPE

winter

SANDERLING *Calidris alba*   to 8"
VOICE: Short *twick* or *quit*.
HABITAT: Outer beaches, tidal flats, lake shores; summer in stony tundra.
NOTES: Looks like a white-bodied, black-legged wind-up toy chasing waves at beach.

BUFF-BREASTED SANDPIPER
Tryngites subruficollis   7½"
VOICE: A low, trilled *pr-r-r-r-reet* or sharp *tik*.
HABITAT: Short-grass prairies, wrack line of beaches, short grass fields, airstrips. Nests in tundra.
NOTES: Migrates mainly through Great Plains; some do pass down the East Coast in fall. Young are tame.

UPLAND SANDPIPER
Bartramia longicauda   to 11½"
VOICE: A mellow, upward-rising, whistled *qwee, qwee, kleweee, cooooleeee*. Calls at night: *kip-ip-ip*.
HABITAT: Grassy prairies, open meadows, fields, airports, golf courses, and sports fields.
NOTES: Alert posture. Often sits atop posts or structures.

PECTORAL SANDPIPER
Calidris melanotos   to 9"
VOICE: Low, reedy, grating *krick*, or *trrip*, when flushed.
HABITAT: Prairie ponds, salt marshes, fresh grassy marshes, muddy shores. Breeds in tundra.
NOTES: Chest stripes that end abruptly. Bulky body. Wears a corduroy bib.

RUFF *Philomachus pugnax*
male to 12", female to 9"
Visitor from Eurasia. Recorded annually in migration. Male is unmistakable in breeding plumage, displaying a large neck ruff and "ear" tufts. Several color forms. Female (Reeve) and winter-plumage males have a thick neck, upright posture, and variable but often yellow-orange legs. Relatively plain face.

SANDPIPERS

summer

winter

SANDERLING

BUFF-BREASTED SANDPIPER

PECTORAL SANDPIPER

breeding dress of ♂ variable

♂

♂

♂

UPLAND SANDPIPER

♂ winter

♀

RUFF

STILT SANDPIPER
Calidris himantopus   **8"**
VOICE: A single, toneless *whu.*
HABITAT: Shallow pools, mud flats, marshes, tundra.
NOTES: A migrant through the Great Plains. Sparse on East Coast, mostly in fall. Mixes with yellowleg or dowitcher flocks. White rump, greenish legs.

CURLEW SANDPIPER
Calidris ferruginea   **to 9"**
Rare straggler to eastern N. America from Eurasia. Occurs with Dunlins and Stilt Sandpipers. White rump.

DUNLIN
Calidris alpina   **to 9"**
VOICE: A nasal, gym whistle–like *treezp.*
HABITAT: Tidal flats, beaches, sandbars, muddy edges of salt marshes. Tundra in summer.
NOTES: Grayish brown in winter with dark down-curved bill. Dark rump. Winters farther north than most shorebirds of the marsh.

PURPLE SANDPIPER
Calidris maritima   **to 9"**
VOICE: A low *weet-wit,* especially when flushed.
HABITAT: Winter: wave-washed rocks and jetties. Breeds in high Arctic.
NOTES: Tame. Yellow legs, 2-toned bill, stocky build. Rare on Great Lakes and Gulf Coast.

SPOTTED SANDPIPER
Actitis macularia   **to 7½"**
VOICE: A clear, high *peet-weet-weet* given when flying, a sharp *weet* when walking.
HABITAT: Pebbly lakeshores, ponds, streams, shorelines, river edges.
NOTES: Bobs (teeters) back end when walking. Stiff-winged, jittery bow-winged flight. Often nests in pebbly areas near beach or parking areas. Scuttles from nest with broken-wing act.

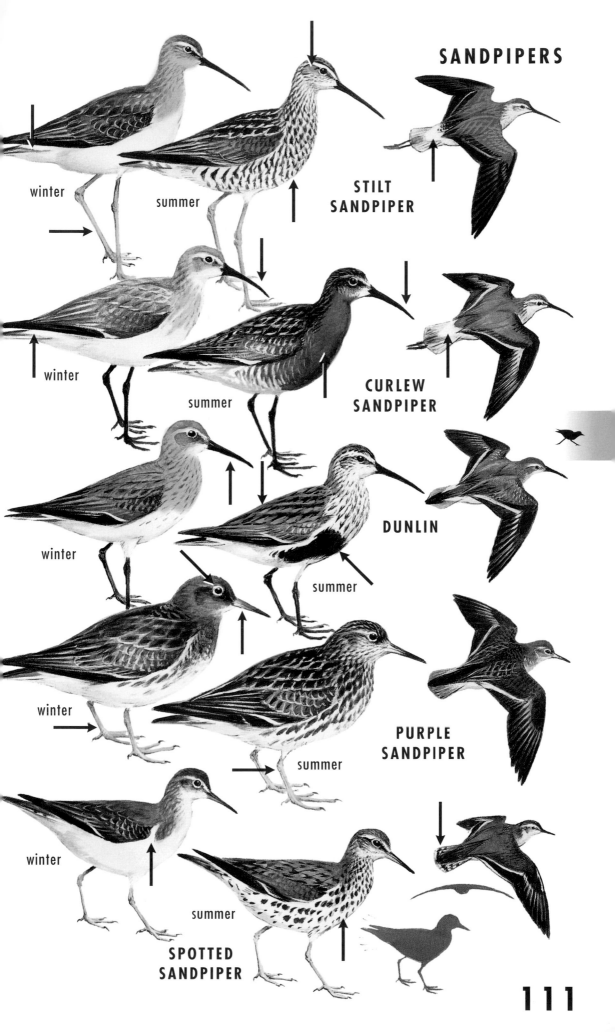

SANDPIPERS

winter

summer

STILT SANDPIPER

winter

summer

CURLEW SANDPIPER

winter

summer

DUNLIN

winter

summer

PURPLE SANDPIPER

winter

summer

SPOTTED SANDPIPER

LEAST SANDPIPER
Calidris minutilla   **to 6½"**
VOICE: A thin *kree-eet*.
HABITAT: Mud flats, grass marshes, rain pools, shores.
NOTES: Our smallest "peep" (collective term for all small sandpipers). Dark brown color, fine bill with slight droop, green-yellow legs. Common.

SEMIPALMATED SANDPIPER
Calidris pusilla   **to 6½"**
VOICE: A *chirt* or *cheh* that lacks *ee* of Least.
HABITAT: Beaches, mud flats; summer in tundra.
NOTES: Most numerous "peep" in migration. Gray-brown with black legs.

WESTERN SANDPIPER
Calidris mauri   **to 7"**
VOICE: Sharp, thin *jeet* or *cheep*.
HABITAT: Shores, beaches, mud flats, inland pond edges.
NOTES: Pale gray above in winter with white on face. Slight droop to long bill. Black legs. To East Coast in fall.

BAIRD'S SANDPIPER
Calidris bairdii   **to 7½"**
VOICE: Trilled *kreeep* or *kree*.
HABITAT: Rain pools, pond margins, mud flats, dry shores, short grass.
NOTES: Long wingtips extend beyond tail. Buffy wash across chest; back looks scaly in immatures. Regular to East Coast only in fall.

WHITE-RUMPED SANDPIPER
Calidris fuscicollis   **to 8"**
VOICE: High, thin *jeet*, like hitting pebbles together.
HABITAT: Prairies, shores, mud flats. Breeds in tundra.
NOTES: The only "peep" with an all-white rump and chevron marks on side below folded wing (breeding adult). Wingtips extend beyond tail end.

NOTE: Comparison of bills of look-alike "peeps" reveals that thickness is important.

"PEEP" SANDPIPERS

LEAST
SANDPIPER

winter   summer

SEMIPALMATED
SANDPIPER

winter   summer

WESTERN
SANDPIPER

winter   summer

BAIRD'S
SANDPIPER

winter   summer

WHITE-RUMPED
SANDPIPER

winter   summer

Least   Semipalmated   Western

BILLS OF SMALL "PEEPS"   ♀

PHALAROPES Family Phalaropodinae

Placed by some taxonomists in the Scolopacidae family. Sandpiper-like birds with lobed toes are at home wading and swimming. When feeding while swimming, they rapidly spin to create a vortex to bring food to surface or rapidly dart about dabbing at surface. Females are more colorful than males, who incubate and raise young.
FOOD: Plankton, brine shrimp, mosquito larvae, and insects.

WILSON'S PHALAROPE
Phalaropus tricolor to 9"
VOICE: Nasal *wurk*, also *check, check, check.*
HABITAT: Shallow prairie lakes, fresh marshes, pools, shores, mud flats. Salt marshes in migration.
NOTES: The most land-loving of the 3 phalaropes. Needlelike bill and no wing marks in flight. Often runs about catching flies at water's edge with tweezerlike grabs.

RED-NECKED PHALAROPE
Phalaropus lobatus to 8"
VOICE: A sharp *kit* or *whit.*
HABITAT: Oceans, bays, lakes, and ponds. Breeds in tundra. During ocean storms, massive numbers (wrecks) can be blown ashore and appear almost anywhere.
NOTES: Of the 2 phalaropes that winter at sea, the Red-necked is most likely to be seen close to shore. Gray winter plumage shows distinct back stripes and dark cap.

RED PHALAROPE
Phalaropus fulicaria to 9"
VOICE: A sharp *whit* or *prip.*
HABITAT: Breeds in tundra but winters mainly at sea.
NOTES: The most pelagic of the phalaropes, the Red has a short, stocky bill and rusty summer plumage. Gray in winter with no back stripes. Scan for it in mats of floating seaweed while at sea. Can be seen along shore after severe storms.

PHALAROPES

winter

♀ summer

winter

winter

WILSON'S PHALAROPE

♂ summer

winter

winter

♀ summer

RED-NECKED PHALAROPE

♂ summer

winter

winter

♀ summer

RED PHALAROPE

♂ summer

lobed foot of
phalarope

LARGE SHOREBIRDS in flight

Learn to know their flight calls, which are quite diagnostic.

HUDSONIAN GODWIT *Limosa haemastica*   p. 104

Upturned bill, white wing stripe, black and white tail. Overhead, black-ish wing linings are seen. Flight call is a *tawit!* (or *godwit!*); it is higher pitched than Marbled's.

WILLET *Catoptrophorus semipalmatus*   p. 106

Contrasting black, gray, and white wing pattern. Overhead, the wing pattern is even more striking. Flight call is a shrill *whee-wee-wee.*

MARBLED GODWIT *Limosa fedoa*   p. 104

Long upturned bill, tawny brown color. Overhead, shows cinnamon wing linings. Flight call is an accented *kerwhit!* (or *godwit!*).

WHIMBREL *Numenius phaeopus*   p. 104

Decurved bill, gray-brown color, striped crown. Overhead, it appears grayer than the Long-billed Curlew; lacks cinnamon wing linings. Flight call is 5–7 short rapid whistles: *ti-ti-ti-ti-ti-ti.*

LONG-BILLED CURLEW *Numenius americanus*   p. 104

Bill is very long and sicklelike; head is not striped. Overhead, shows cinnamon wing linings. Flight call is a rapid, whistled *kli-li-li-li* or *cur-lee.*

HUDSONIAN
GODWIT
summer

WADERS
in flight

HUDSONIAN
GODWIT
winter

WILLET
winter

MARBLED
GODWIT

WHIMBREL

LONG-BILLED
CURLEW

117

SNIPE, SANDPIPERS, etc. in flight

These and the shorebirds on the next plate are figured in black and white to emphasize their basic flight patterns. Most of these have unpatterned wings, lacking a central stripe. All are shown in full color on previous plates. Learn their distinctive flight calls.

COMMON SNIPE *Gallinago gallinago*   **p. 102**
Flight note, when flushed, is a rasping *scaip*.

AMERICAN WOODCOCK *Scolopax minor*   **p. 102**
Wings whistle in flight. At night, an aerial flight "song."

SOLITARY SANDPIPER *Tringa solitaria*   **p. 106**
Flight note, *peet!* or *peet-weet!* (higher than Spotted's).

LESSER YELLOWLEGS *Tringa flavipes*   **p. 106**
Flight call, *yew* or *yu-yu*, softer than Greater's.

GREATER YELLOWLEGS *Tringa melanoleuca*   **p. 106**
Flight call, a forceful 3-note whistle, *whew-whew-whew!*

WILSON'S PHALAROPE *Phalaropus tricolor*   **p. 114**
Flight note is a low, nasal *wurk*.

STILT SANDPIPER *Calidris himantopus*   **p. 110**
Flight note is a single *whu*, lower than Lesser Yellowlegs'.

UPLAND SANDPIPER *Bartramia longicauda*   **p. 108**
Flight note is a mellow, whistled *kip-ip-ip-ip*.

BUFF-BREASTED SANDPIPER *Tryngites subruficollis*   **p. 108**
Flight note is a low, trilled *pr-r-r-reet*.

PECTORAL SANDPIPER *Calidris melanotos*   **p. 108**
Flight note is a low, reedy *krik* or *krik-krik*.

WADERS in flight

COMMON SNIPE

AMERICAN WOODCOCK

SOLITARY SANDPIPER

GREATER YELLOWLEGS

LESSER YELLOWLEGS

WILSON'S PHALAROPE
winter

STILT SANDPIPER
winter

UPLAND SANDPIPER

BUFF-BREASTED SANDPIPER

PECTORAL SANDPIPER

SANDPIPERS, PHALAROPES in flight

Most of these have a light wing stripe. Learn their calls.

DOWITCHERS *Limnodromus* ssp. **p. 102**
Flight call of Short-billed Dowitcher (*L. griseus*), a staccato trebled *tututu*; that of Long-billed Dowitcher (*L. scolopaceus*), a single thin *keek*, occasionally doubled or trebled.

DUNLIN *Calidris alpina* **p. 110**
Flight note is a nasal rasping *cheezp* or *treezp*.

RED KNOT *Calidris canutus* **p. 102**
Flight note is a low *knut*.

PURPLE SANDPIPER *Calidris maritima* **p. 110**
Flight note is a low *weet-wit* or *twit*.

WHITE-RUMPED SANDPIPER *Calidris fuscicollis* **p. 112**
Flight note is a high mouselike squeak, *jeet*.

CURLEW SANDPIPER *Calidris ferruginea* **p. 110**

RUFF *Philomachus pugnax* **p. 108**

SPOTTED SANDPIPER *Actitis macularia* **p. 110**
Flight note is a clear *peet* or *peet-weet*.

SANDERLING *Calidris alba* **p. 108**
Flight note, a short *twick* or *quit*.

RED PHALAROPE *Phalaropus fulicaria* **p. 114**

RED-NECKED PHALAROPE *Phalaropus lobatus* **p. 114**
Flight note (both phalaropes) is a sharp *whit* or *kip*.

LEAST SANDPIPER *Calidris minutilla* **p. 112**
Flight note is a thin *kree-eet*.

SEMIPALMATED SANDPIPER *Calidris pusilla* **p. 112**
Flight note is a soft *chet* or *chirt* (lacks *ee* sound of Least).

WESTERN SANDPIPER *Calidris mauri* (not shown) **p. 112**
Flight note is a thin *jeet* or *cheep*.

BAIRD'S SANDPIPER *Calidris bairdii* **p. 112**
Flight note is a low, rough *kreep* or *kree*.

WADERS in flight

SHORT-BILLED
DOWITCHER
(Long-billed has
similar pattern)

winter

RED
KNOT

winter

DUNLIN

PURPLE
SANDPIPER

WHITE-RUMPED
SANDPIPER

CURLEW
SANDPIPER
winter

winter

RUFF

winter

SPOTTED
SANDPIPER

SANDERLING

winter

RED
PHALAROPE

winter

RED-NECKED
PHALAROPE

LEAST
SANDPIPER

SEMIPALMATED
SANDPIPER

BAIRD'S
SANDPIPER

121

TURKEYS Family Meleagrididae

Large fowl. Iridescent feathering, naked head. Male has a "beard" and fans tail in display. **FOOD:** Berries, acorns, nuts, seeds, insects.

WILD TURKEY *Meleagris gallopavo*
male to 48", female to 36"
VOICE: A rolling gobble or an alarm *pit*. Female: *keow-keow*.
HABITAT: Woods, mountain forests, swamplands, suburbs.
NOTES: One of our most recognizable birds. Populations have increased dramatically in response to restocking efforts. Found locally north to dash line.

GROUSE Family Tetraonidae

Gound-dwelling chickenlike birds lacking long tail of pheasants. Explode into flight when flushed. **FOOD:** Seeds, buds, berries, insects.

RUFFED GROUSE *Bonasa umbellus* **to 19"**
VOICE: Male creates sound with his wings, a low booming that accelerates into a whir: *bup-bup-bup-bupbupbupup-r-r-rrrrrr*.
HABITAT: Understory of deciduous or mixed woodlands.
NOTES: Heard more than seen. Spends time in treetops eating buds in spring. Red and gray forms.

PHEASANTS Family Phasianidae (in part)

Chicken-sized ground-dwelling birds. Reluctant to fly. **FOOD:** Seeds, berries, buds, insects.

RING-NECKED PHEASANT
Phasianus colchicus
male to 36" (including tail), female to 25"
VOICE: Male: a harsh crowing *kork-kork*. Female: a low clucking.
HABITAT: Farms, fields, brush and marsh edges.
NOTES: Introduced. Male has a long tail and red face wattles. Female lacks any adornment. This ground nester flushes with an explosive croak.

MISCELLANEOUS FOWLLIKE BIRDS

display

♂

♂

♀

WILD TURKEY

♂

♂

RUFFED GROUSE

gray form

red form

♂

♂

display

♂

♀

♂

♀

RING-NECKED PHEASANT

123

SPRUCE GROUSE
Falcipennis canadensis　　　　　　**to 17"**
VOICE: A low, churring cluck.
HABITAT: Evergreen forests, muskeg, berry patches.
NOTES: A very tame, dark grouse. Chestnut band on tail tip in both male and female. A skulker of the dark understory.

SHARP-TAILED GROUSE
Tympanuchus phasianellus　　　　**to 20"**
VOICE: Courting: a low *coo-oo* and buzzing vibration of tail. Also a cackling *cac-cac-cac*.
HABITAT: Prairies, open fields, forest edges, clearings.
NOTES: Elongate central tail feathers with white outer portion. Male displays in open areas with inflated pink neck pouch, foot shuffling, and tail quill rattling.

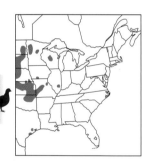

GREATER PRAIRIE-CHICKEN
Tympanuchus cupido　　　　　　**to 18"**
VOICE: Clucking. In breeding season male displays producing a booming sound from neck sack, a hollow *oo-loo-woo*.
HABITAT: Tall-grass prairies at limited sites.
NOTES: Courtship on century-old dancing grounds. Male struts, inflates yellow-orange neck pouch, and raises black neck feathers.

LESSER PRAIRIE-CHICKEN
Tympanuchus pallidicinctus　　　**to 16"**
VOICE: Clucking. In courtship, the Lesser sounds a lower booming than the Greater that doesn't carry as far.
HABITAT: Sandhill country of sage, bluestem grass, and scrub oak.
NOTES: Courtship display similar to Greater's. The Lesser's airsacs are plum-colored. More western. Limited habitat and display areas.

GROUSE

SPRUCE GROUSE

♀

♂

SHARP-TAILED GROUSE

♂

♂ display

GREATER PRAIRIE-CHICKEN

♂

♂ display

LESSER PRAIRIE-CHICKEN

♂

♂ display

125

WILLOW PTARMIGAN

Lagopus lagopus   **to 16"**

VOICE: Deep, raucous, almost humanlike calls: *puck puck puck-pud-errr,* then *go back-go back* or *to-bacco, tobacco.*

HABITAT: Timberline in mountains, willow scrub, and tundra.

NOTES: Changes colors by season. Remarkable camouflage. White in front of eye in winter. In spring it is a richer chestnut color with larger bill than Rock Ptarmigan's. Wanders irregularly to dash line.

ROCK PTARMIGAN *Lagopus mutus*   **to 13"**

VOICE: Clucks and sounds a rapid *put-put-put-dii-derrrr,* especially when flushed.

HABITAT: Rocky outcrops of alpine areas. Tends to live in higher areas than the Willow. Tundra.

NOTES: Brown color and black line in front of eye in winter plumage. Rare in winter to dash line.

QUAILS and PARTRIDGES
Family Phasianidae (in part)

GRAY PARTRIDGE *Perdix perdix*   **to 14"**

VOICE: *Karr-wit karr-wit.*

NOTES: Introduced from Eurasia as a game bird. Has done well in some areas but disappeared from others. Chestnut face and underbelly. Plump.

NORTHERN BOBWHITE

Colinus virginianus   **to 10½"**

VOICE: Very familiar, whistled *bob-white* with upward inflection on second note. Covey call is a sharp *whoil-eeeek.*

HABITAT: Farms, brushy open country, wood edges.

NOTES: Small, rotund. Heard more often than seen. Explodes into flight when approached too closely. Numbers renewed due to restocking in North.

SCALED QUAIL *Callipepla squamata* **to 12"**
A scaly, white "cotton-topped" quail. Western.

PTARMIGANS

♀ summer

winter

♂ summer

WILLOW PTARMIGAN

♀ summer

winter

♂ summer

ROCK PTARMIGAN

QUAIL, etc.

sexes similar

GRAY PARTRIDGE

SCALED QUAIL

NORTHERN BOBWHITE

♂

♀

127

HAWKS, EAGLES, etc. Family Accipitridae

Diurnal (daytime) birds of prey with hooked bills and sharp talons. Very important to the ecosystem and afforded protection after years of persecution. Several subfamilies.

KITES Subfamilies Elaninae and Milvinae

These graceful birds of prey of southern and western distribution occasionally stray into the Northeast. FOOD: Large insects, snakes, rodents, invertebrates.

SWALLOW-TAILED KITE
Elanoides forficatus   24"
VOICE: A shrill *ee-ee-ee* or *pee-pee-pee*.
HABITAT: Wooded river swamps and pine woods.
NOTES: Striking black and white with deeply forked tail. Graceful flight. Spring overshoots to North.

MISSISSIPPI KITE
Ictinia mississippiensis   14"
VOICE: A *phee-phew* (Sutton) or a clear *kee-ee*.
HABITAT: Wooded streams, groves, shelterbelts.
NOTES: Soft gray bird with black eye smudge. Black tail flares at tip. Range appears to be spreading.

WHITE-TAILED KITE
Elanus leucurus   to 17"
VOICE: Mainly silent but gives thin, whistled *see* or *see-see*.
HABITAT: Open groves, river valleys, marshes.
NOTES: Hovers when hunting. White in appearance with black in wing.

SNAIL KITE (Everglades Kite)
Rostrhamus sociabilis   to 19"
VOICE: Crackling *kor-ee-ee-a, kor-ee-ee-a*.
HABITAT: Fresh marshes and canals with *Pomacea* snails (apple snails).
NOTES: Snail Kites are restricted to Florida in the U.S. Population fluctuates depending on water levels in glades and canals and on apple snail population.

Overhead flight patterns on p. 149

KITES

SWALLOW-TAILED KITE

MISSISSIPPI KITE

immature

adult

adult

adult

immature

WHITE-TAILED KITE

immature

adult

adult

immature

♀

♂

SNAIL KITE (Everglades Kite)

♂

129

ACCIPITERS (BIRD HAWKS) Subfamily Accipitrinae

These long-tailed woodland hawks have short, rounded wings. Typical flight pattern is several beats and a glide. Females are larger. FOOD: Chiefly birds with some small mammals.

SHARP-SHINNED HAWK
Accipiter striatus   to 14"
VOICE: A high *kik-kik-kik*.
HABITAT: Woods and thickets.
NOTES: Trim body, square tail, and small head. Fast, snappy wingbeats before glide. Very local breeder in the South.

COOPER'S HAWK *Accipiter cooperii* to 20"
VOICE: Rapid *kek, kek, kek* at nest site.
HABITAT: Broken woodlands, and river groves.
NOTES: Has a rounded tail tip, large head, and longer neck and straight-edged leading edge to wing compared to Sharp-shinned. Like Sharp-shinned, a regular at birdfeeding stations.

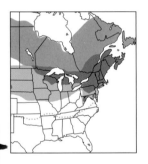

NORTHERN GOSHAWK
Accipiter gentilis   to 26"
VOICE: A sharp, loud *kak, kak, kak* or *kuk, kuk, kuk*. Tenaciously defends nest site.
HABITAT: Evergreen forests and deciduous woods.
NOTES: Cyclic winter visitor south to dash line.

HARRIERS Subfamily Circinae

Slim hawks with slim wings and long tails. Flight is low and languid. Glide with wings in a dihedral (shallow V). Sexes are not alike.

NORTHERN HARRIER
Circus cyaneus   to 24"
VOICE: A weak, nasal, pleading *pee, pee, pee*.
HABITAT: Open country. Fields, marshes, prairies, muskeg.
NOTES: White rump and wing position diagnostic. Male gray, female and immature brown. Populations have suffered because of habitat loss.

ACCIPITERS, HARRIERS

ACCIPITERS
(True Hawks)
have small heads,
short, rounded wings,
and long tails

immature

adult

SHARP-SHINNED HAWK

immature

adult

COOPER'S HAWK

immature

adult

NORTHERN GOSHAWK

Northern Harrier

immature

♂

♀

NORTHERN HARRIER

131

BUTEOS (BUZZARD HAWKS)
Subfamily Buteoninae (in part)

Large, thickset hawks with broad wings. Habitually soar high and in wide circles. Migrant groups often form kettles of birds on thermals (rising warm air masses). **FOOD:** Rodents, rabbits; occasional birds, reptiles, insects.

RED-TAILED HAWK

Buteo jamaicensis   **to 25"**

VOICE: Downward-slurred scream: *keeer-eee.*
HABITAT: Open country, woods, prairie groves, mountains, plains.
NOTES: Common roadside hawk. Pale patches on back. Tail orange to orange-buff in adult, brown in immature. Dark belly band, hooded head.

"KRIDER'S" RED-TAILED HAWK

Buteo jamaicensis kriderii

Pale prairie race with white tail tipped rufous. Look for in prairie provinces of Canada and plains states.

"HARLAN'S" RED-TAILED HAWK

Buteo jamaicensis harlani

Variable black race of Red-tail; formerly regarded as distinct species. Very dark with pale tail with dark mottling.

SWAINSON'S HAWK

Buteo swainsoni   **to 22"**

VOICE: A shrill, plaintive whistle: *kreeeee.*
HABITAT: Plains, rangeland, open hills, sparse trees.
NOTES: Dark-chested with pale throat. Glides with wings slightly above horizontal. Flight feathers of wings are dark. Rare to East, especially in fall.

FERRUGINOUS HAWK *Buteo regalis* **to 25"**

VOICE: Screams and cackles.
HABITAT: Plains, savannahs, rocky canyons.
NOTES: Straggler east of Mississippi River. Rufous thighs.

BUTEOS have
stocky build,
wide tail

immature

adult

RED-TAILED
HAWK

BUTEOS

Overhead patterns
on pp. 143, 147

"HARLAN'S"
RED-TAILED
HAWK

"Harlan's"
Red-tailed

"KRIDER'S"
RED-TAILED
HAWK

immature

SWAINSON'S
HAWK

dark
form

light
form

FERRUGINOUS
HAWK

133

ROUGH-LEGGED HAWK
Buteo lagopus   to 24"
VOICE: A pleading *kaaarrr.*
HABITAT: Summer: tundra; winter: open plains, coastal marshes, fields.
NOTES: Large hawk of open country. Hovers while hunting. Dark and light forms; both adults and immature have a dark wrist patch. Tail is white with dark terminal band. Winter numbers vary.

RED-SHOULDERED HAWK
Buteo lineatus   to 24"
VOICE: A 2-syllabled *kee-yer* that drops in inflection.
HABITAT: Timbered swamps, woodlands, river edge.
NOTES: In flight, translucent crescent patch shows at base of primaries. Florida race pale and tame. Immature streaked below and fairly similar to Broad-winged.

BROAD-WINGED HAWK
Buteo platypterus   to 19"
VOICE: A high-pitched, shrill 2-note *p-weeeee.*
HABITAT: Woods, groves.
NOTES: Small buteo, adult with black and white tail banding. Many migrants soar in kettles.

SHORT-TAILED HAWK
Buteo brachyurus   17"
VOICE: A shrill, piercing *shreeeea,* also soft *kleeee* near nest.
HABITAT: Pine woods, wood edges, cypress swamps, mangrove swamps.
NOTES: The size of a crow. Dark and light morph; light morph is pure white below and dark morph is fully dark. Banded tail. Concentrate in south Florida in winter.

dark form

light form

immature

ROUGH-LEGGED HAWK

immature

adult

RED-SHOULDERED HAWK

immature

pale
s. Florida
form

immature

adult

adult

immature

BROAD-WINGED HAWK

light
morph
immature

dark form

light form

light form
adult

SHORT-TAILED HAWK

dark form

135

EAGLES Subfamily Buteoninae (in part)

Eagles are distinguished from buteos by their much larger size and proportionately longer wings. **FOOD:** Mainly fish; also carrion.

BALD EAGLE *Haliaeetus leucocephalus*
to 43" (wingspan to 8')
VOICE: A thin *kleek-kik-kik-kik*.
HABITAT: Coastline, offshore islands, large lakes, and rivers; mountain ridges in migration.
NOTES: The national bird of the U.S. First-year birds chocolate brown, then taking up to 5 years to attain white head and tail. Build massive stick nests. Scattered pairs nest south to dash line.

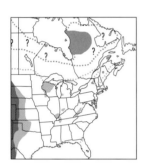

GOLDEN EAGLE *Aquila chrysaetos*
40" (wingspan to 7')
VOICE: A yelping *kya-kya*, high whistles.
HABITAT: Open mountains, foothills, plains, open country. Woodlands, river systems, and marshes in winter.
NOTES: This massive eagle holds its broad wings flat or in a slight V. Adult has a golden nape. Immature birds have a whitish tail base, dark tail tip, and white patches at base of primaries. Little is known about its e. Canada breeding range.

OSPREYS Family Pandioninae

One species is found worldwide. Only raptor that plunges into water for fish. Sexes alike. Builds massive stick nests. **FOOD:** Fish.

OSPREY *Pandion haliaetus*
to 24½" (wingspan to 6')
VOICE: A high-pitched, pleading *cheep, cheep cheep* or *pleep, pleep, pleep*. At nest a frenzied *cheeerk*.
HABITAT: Rivers, lakes, coastline.
NOTES: Do not confuse white head and black mask with subadult Bald Eagle. Distinct crook to wing in flight. Has taken to nest platforms erected in many areas. Scattered pairs nest south of mapped range.

EAGLES

BALD
EAGLE
adult

Overhead
patterns
on p. 145

BALD EAGLE
immature

GOLDEN
EAGLE
adult

GOLDEN
EAGLE
immature

OSPREY

hovering

OSPREY

adult

137

AMERICAN VULTURES Family Cathartidae

Eaglelike birds with small naked heads. They are often seen soaring in wide circles. Incorrectly called buzzards. FOOD: Primarily carrion.

TURKEY VULTURE *Cathartes aura*
32" (wingspan 6')

VOICE: Usually silent. Hisses loudly at nest.

HABITAT: Soars over forests, cliffs, plains.

NOTES: Holds wings in a dihedral (shallow V); rocks back and forth while soaring. Young with gray head.

BLACK VULTURE *Coragyps atratus*
to 27" (wingspan under 5')

VOICE: Usually silent; hisses, grunts at nest and while feeding in flocks.

HABITAT: Similar to Turkey Vultures, the Black Vulture tends to avoid higher elevations.

NOTES: Flaps rapidly while flying. Has distinct white wing patches and a very short tail. Spreading northward.

CARACARAS and FALCONS Family Falconidae

CARACARAS Subfamily Caracarinae

Large, long-legged birds of prey with naked faces. Aberrant group of falcons. Sexes alike. FOOD: Snakes, insects, lizards.

CRESTED CARACARA
Caracara plancus   to 25"

VOICE: A harsh, cackling *kerr-kerr-kerr*; deep growling at nest site.

HABITAT: Prairies and rangelands.

NOTES: Spectacular and long-legged with crest and red face. White wing patches. Rowing flight.

SCAVENGERS

adult

immature

TURKEY
VULTURE

BLACK
VULTURE

adult

mature

adult

CRESTED
CARACARA

above, King Vulture
(formerly in Florida)

FALCONS Subfamily Falconinae

Streamlined birds of prey with pointed wings and longish tails. Swift in flight. FOOD: Birds, rodents, insects.

AMERICAN KESTREL
Falco sparverius   **to 12"**
VOICE: A rapid, high-pitched *klee-klee-klee*.
HABITAT: Open country, farmland, fields, wood edges, and even cities. Roadside wires and posts.
NOTES: Hovers while hunting. Nests in holes, trees, and boxes set out. Common migrant.

MERLIN
Falco columbarius   **to 13½"**
VOICE: *Kee, kee, kee,* mainly on breeding ground.
HABITAT: Open evergreen woods. In migration open country, coasts, foothills.
NOTES: Has banded tail and facial mustache.

PRAIRIE FALCON *Falco mexicanus*   **to 17"**
A rare wanderer to the eastern prairie states. Western species resembles immature Peregrine. (See p. 149.)

PEREGRINE FALCON
Falco peregrinus   **to 20"**
VOICE: At nest: *we-chew*, also *kek-kek-kek*.
HABITAT: Open country, from mountains to coast. Found worldwide.
NOTES: A large, powerful falcon with broad wings and a heavy facial mustache. Reintroduced to many areas, including cities south of Arctic breeding range to dash line. Many are resident.

GYRFALCON *Falco rusticolus*   **to 25"**
VOICE: A loud, sharp kacking near nest site.
HABITAT: Arctic tundra, cliffs, seacoasts.
NOTES: Has 3 color morphs. Larger than a Peregrine with a slower wingbeat and broader wings and tail. Rarely travels to dash line in winter.

FALCONS

Overhead patterns on p. 149

FALCONS have long, pointed wings, long tails

♀ ♂

Kestrel

AMERICAN KESTREL

♀

MERLIN

♂ ♂

Merlin

Peregrine

immature

adult

PRAIRIE FALCON

PEREGRINE FALCON

gray morph

white morph

dark morph

GYRFALCON

BUTEOS AND HARRIERS OVERHEAD

The birds shown opposite are mostly adults.

 **BUTEOS** are chunky, with broad wings and broad, rounded tails. They soar and wheel high in the open sky.

RED-TAILED HAWK *Buteo jamaicensis*   p. 132
Light chest, streaked belly. Dark patch on leading edge of underwing. Tail has little or no banding.

RED-SHOULDERED HAWK *Buteo lineatus*   p. 134
Banded tail (white bands narrow). Translucent crescent-shaped wing "windows."

SWAINSON'S HAWK *Buteo swainsoni*   p. 132
Dark chest, light wing linings, dark flight feathers. Immature has streaked breast.

BROAD-WINGED HAWK *Buteo platypterus*   p. 134
ADULT: Broadly banded tail (white bands wide); whitish underwing .
IMMATURE: Narrowly banded tail; whitish underwing.

SHORT-TAILED HAWK *Buteo brachyurus*   p. 134
LIGHT MORPH: Clear white belly and wing linings. (Only Florida buteo so colored.)

ROUGH-LEGGED HAWK *Buteo lagopus*   p. 134
LIGHT MORPH: Dark belly, black wrist patch. Whitish tail with broad dark band.

FERRUGINOUS HAWK *Buteo regalis*   p. 132
White or pale rusty tail; dark V formed by rufous legs (adult only). Whitish at base of primaries on upperwing.

 **HARRIERS** are slim with long, rounded wings and long tails. They fly low with a vulturelike dihedral.

NORTHERN HARRIER *Circus cyaneus*   p. 130
MALE: Slim; pale with black wingtips. **FEMALE:** Harrier shape; brown; streaked and barred tail.

142

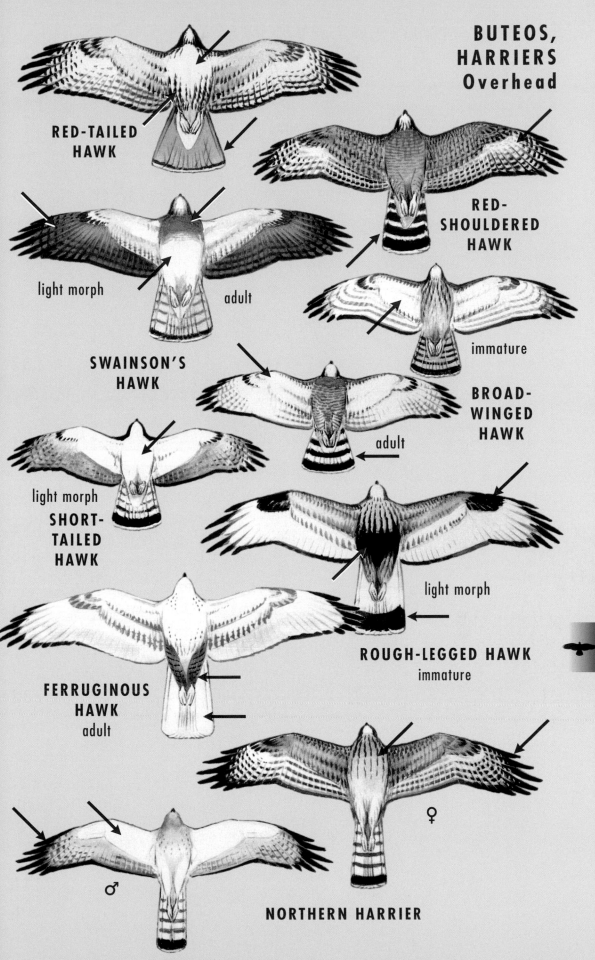

**RED-TAILED
HAWK**

**RED-
SHOULDERED
HAWK**

light morph

adult

**SWAINSON'S
HAWK**

immature

**BROAD-
WINGED
HAWK**

adult

light morph

**SHORT-
TAILED
HAWK**

light morph

ROUGH-LEGGED HAWK
immature

**FERRUGINOUS
HAWK**
adult

♀

♂

NORTHERN HARRIER

143

BALD EAGLE *Haliaeetus leucocephalus*   **p. 136**
ADULT: White head and tail. **IMMATURE:** Some white in wing linings.

GOLDEN EAGLE *Aquila chrysaetos*   **p. 136**
ADULT: Almost uniformly dark; wing linings dark. **IMMATURE:** "Ringed" tail; white patches at base of primaries.

OSPREY *Pandion haliaetus*   **p. 136**
Clear white belly; black wrist patches.

Where the **BALD EAGLE, TURKEY VULTURE,** and **OSPREY** are all found, they can be separated at a great distance by their manner of soaring: the Bald Eagle soars with flat wings; the Turkey Vulture soars with a dihedral; the Osprey often soars with a kink or crook in its wings.

BALD EAGLE   adult

BALD EAGLE   immature

GOLDEN EAGLE   adult

GOLDEN EAGLE   immature

OSPREY

BLACKISH BIRDS OF PREY OVERHEAD

TURKEY VULTURE *Cathartes aura* **p. 138**
2-toned wings, small head, longish tail.

BLACK VULTURE *Coragyps atratus* **p. 138**
Stubby tail, white wing patches.

CRESTED CARACARA *Caracara plancus* **p. 138**
Whitish chest, black belly, pale patches on primaries.

SNAIL KITE *Rostrhamus sociabilis* **p. 128**
IMMATURE: Brown, streaked; white tail with black band. **ADULT MALE:** Dark slate body, white tail with black band.

ROUGH-LEGGED HAWK *Buteo lagopus* **p. 134**
DARK MORPH: Dark body, whitish flight feathers; whitish tail with wide black terminal band.

SWAINSON'S HAWK *Buteo swainsoni* **p. 132**
DARK MORPH: Wings are usually dark throughout, including flight feathers. Wings narrow and pointed for buteo; held in slight V. Tail is narrowly banded.

"HARLAN'S" RED-TAILED HAWK
Buteo jamaicensis harlani **p. 132**
Red-tailed Hawk shape, dark body; pale, mottled tail. Other melanistic Red-tails may be similar except for tail.

SHORT-TAILED HAWK *Buteo brachyurus* **p. 134**
DARK MORPH: Black body and wing linings, pale banded tail. The only small *blackish* buteo in Florida.

DARK BIRDS OF PREY
Overhead

TURKEY
VULTURE

BLACK
VULTURE

CRESTED
CARACARA

SNAIL KITE
immature

SNAIL KITE
♂

ROUGH-LEGGED
HAWK
dark morph

SWAINSON'S
HAWK
dark morph

"HARLAN'S"
RED-TAILED
HAWK

dark morph

SHORT-TAILED HAWK

147

ACCIPITERS, FALCONS, AND KITES OVERHEAD

COOPER'S HAWK *Accipiter cooperii*   **p. 130**
Underparts rusty (adult). Near size of American Crow. Tail of female is rounded, even when folded; tail of male less so. Longer head projection.

NORTHERN GOSHAWK *Accipiter gentilis*   **p. 130**
ADULT: Larger than American Crow; pale gray underbody (adult). Inner-wing bulges.

SHARP-SHINNED HAWK *Accipiter striatus*   **p. 130**
Jay-sized or a bit larger; tail of male is square or notched when folded; tail of female less so. Fanned tail will seem more rounded. Shorter head projection.

GYRFALCON *Falco rusticolus*   **p. 140**
DARK MORPH: Blacker below than Peregrine. **GRAY MORPH:** More uniformly colored than Peregrine. **WHITE MORPH (NOT SHOWN):** White as a Snowy Owl.

AMERICAN KESTREL
Falco sparverius   **p. 140**
Small; rufous tail has a dark tip or dark bands.

MERLIN *Falco columbarius*   **p. 140**
Small; dark; banded gray tail. Streaked underparts.

PEREGRINE FALCON *Falco peregrinus*   **p. 140**
Falcon-shaped; size near that of American Crow; bold face pattern.

PRAIRIE FALCON *Falco mexicanus*   **p. 140**
Size of Peregrine; paler; note black patch in wing linings.

SWALLOW-TAILED KITE *Elanoides forficatus*   **p. 128**
White body and wing linings; deeply forked black tail.

WHITE-TAILED KITE *Elanus leucurus*   **p. 128**
Falcon-shaped; white body, whitish tail.

MISSISSIPPI KITE *Ictinia mississippiensis*   **p. 128**
Falcon-shaped. **ADULT:** Black tail, blackish wings, gray body. **IMMATURE:** Streaked breast; banded square-tipped or notched tail.

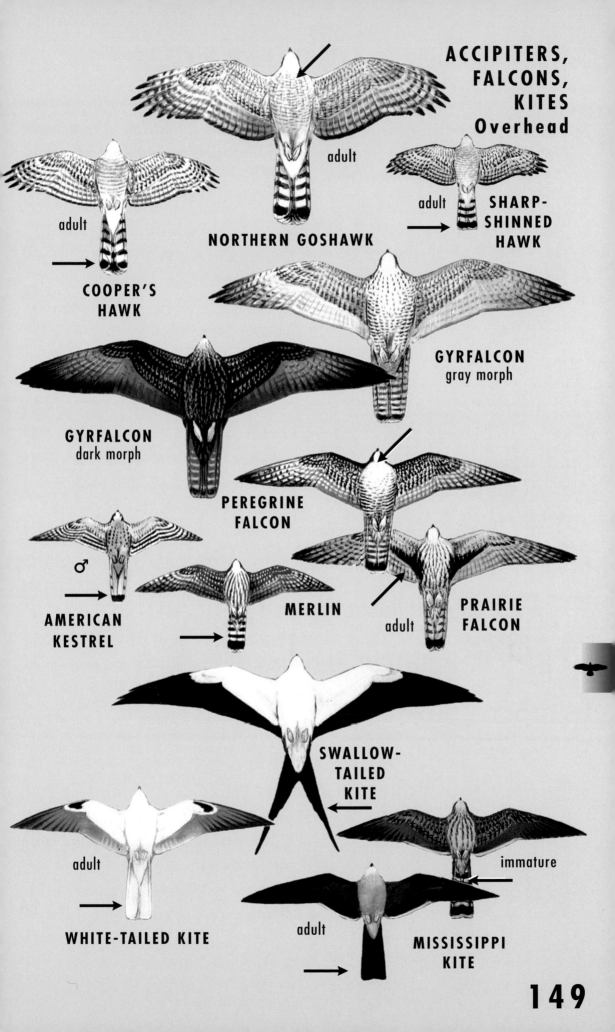

ACCIPITERS, FALCONS, KITES
Overhead

adult

NORTHERN GOSHAWK

adult

SHARP-SHINNED HAWK

adult

COOPER'S HAWK

GYRFALCON
gray morph

GYRFALCON
dark morph

PEREGRINE FALCON

♂

AMERICAN KESTREL

MERLIN

adult

PRAIRIE FALCON

SWALLOW-TAILED KITE

adult

WHITE-TAILED KITE

immature

adult

MISSISSIPPI KITE

149

OWLS Families Tytonidae (Barn Owls) and Strigidae (True Owls)

Chiefly nocturnal birds of prey with large heads, flattened faces that form facial disks, and large, forward-facing eyes. Hooked bills and claws. Flight is noiseless and mothlike. **FOOD:** Rodents, birds, reptiles, fish, insects.

SHORT-EARED OWL *Asio flammeus*   to 17"
VOICE: Doglike *yap-yap-yap* or *waow*.
HABITAT: Prairies, marshes, dunes, tundra.
NOTES: Owl of open country. Hunts in evening, early morning, and sometimes during the day, especially when overcast. Buffy wing patches; black wrist patches on underwing. Population is cyclic. Breeds rarely and locally south to dash line.

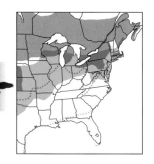

EASTERN SCREECH-OWL *Otus asio*   to 10"
VOICE: A mournful whinny or wail with a tremulous descending pitch or a pulsing single-pitch series.
HABITAT: Woodlands, farms, parks, towns.
NOTES: Abundant small owl with ear tufts. Red and gray morphs. Nests in holes in trees, where it may sit in morning or winter daylight.

LONG-EARED OWL *Asio otus*   to 16"
VOICE: A moaning *whooooo* or a doglike barking.
HABITAT: Woodlands, thickets, conifer groves. Local.
NOTES: A slender, crow-sized owl with closely set ear tufts and large orange facial disks. Freezes in position close to trunk. In winter, often roosts in groups Breeds and winters irregularly south to 2 dash lines.

GREAT HORNED OWL
Bubo virginianus   to 25"
VOICE: A distinct, low, 5-parted *hoo, hoo-oo hoo, hoo*. Female's call is lower, with up to 8 hoots.
HABITAT: Forest, woodlands, thickets, farmlands, in parks and suburban areas. Uses old crow andhawk nests.
NOTES: Our largest, most powerful owl. Has ear tufts, large yellow eyes, and a white throat.

SHORT-EARED OWL

"EARED" OWLS

rufous morph

EASTERN SCREECH-OWL

gray morph

LONG-EARED OWL

subarctic morph

GREAT HORNED OWL

typical form

151

BARRED OWL *Strix varia*   to 24"

VOICE: Distinctive. Eight accented hoots: *who-cooks-for-you-who-cooks-for-you'all.*

HABITAT: River bottoms, swamps, woodlands.

NOTES: Large; puffy-headed. Throat bars crosswise and streaked lengthwise on belly. Brown eyes. Nests in old crow nests or tree cavities.

BARN OWL *Tyto alba*   to 20"

VOICE: A sharp, rasping hiss: *shiiiish.*

HABITAT: Open country. Nests in tree cavities, old abandoned farm and town buildings, cliffs, under trestles; may roost in conifer groves.

NOTES: Golden white with heart-shaped facial disk. Mothlike flight. More common in South.

GREAT GRAY OWL *Strix nebulosa*   to 33"

VOICE: A deep, booming *whoo.*

HABITAT: Conifer and aspen woods and adjacent meadows and bogs.

NOTES: A large owl. Facial disk is large with concentric circles; it has piercing yellow eyes. It is rare in most of U.S.; does have invasion years east and south to dash line.

SNOWY OWL *Nyctea scandiaca*   to 27"

VOICE: Usually silent. Deep *whoo* and clicks near nest.

HABITAT: Open areas, rooftops, coastal dunes, fields. Nests circumpolar in Arctic. Invades northern portions of U.S. sporadically in winter south to dash line.

NOTES: Male is pure white; female and immature are heavily barred.

BARRED OWL

BARN OWL

GREAT GRAY OWL

SNOWY OWL

BOREAL OWL *Aegolius funereus*   to 10"

VOICE: A staccato, winnowing note. A pulsating *whew-whew-whew* that descends slightly.

HABITAT: Mixed-wood and conifer forests.

NOTES: Has a small squarish face with black rim, a spotted forehead, and a yellow bill. Rare visitor to northern states south to dash line. Usually very tame.

NORTHERN SAW-WHET OWL

Aegolius acadicus   to 8½"

VOICE: A repetitive, whistled *too-too-too-too-too*, up to 130 times per minute.

HABITAT: Forests, conifers, swamplands.

NOTES: Has a very small rounded head, streaked forehead, and black bill. Tame. Breeds and winters irregularly south to dash lines.

BURROWING OWL

Athene cunicularia   to 11"

VOICE: A rapid, chattering *quick-quick-quick*. At night near burrow: a dovelike cooing.

HABITAT: Open grasslands, prairies, farmlands, airfields, even lawns in towns.

NOTES: Small. Active in daytime, it is long-legged and runs about bent over. Sits atop posts or berm of burrow.

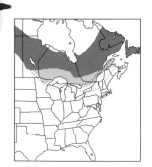

NORTHERN HAWK OWL

Surnia ulula   to 17½"

VOICE: A chattering *kikikiki* or kestrellike *killy-illy-illy.*

HABITAT: Boreal forests, birch scrub, bogs.

NOTES: A rare visitor from the North. Day-flying. Tame. It looks more like a hawk than an owl; it is gray with a long tail; flat face is outlined in black. Flies low to ground and swoops up to perch.

BOREAL OWL

NORTHERN
SAW-WHET OWL
juvenile

NORTHERN
SAW-WHET OWL
adult

NORTHERN HAWK OWL

BURROWING OWL

PARROTS, PARAKEETS Family Psittacidae

Compact, short-necked birds with stout, hooked bills. Noisy. Carolina Parakeet was the only native parakeet to occur in e. U.S. Now extinct, it was last reported in 1920 in Florida. A number of exotic species have been released or escaped; several species have established themselves. **FOOD:** Fruits and seeds.

MONK PARAKEET *Myiopsitta monachus* 11½"
VOICE: Raucous squawking.

Long tail and gray hood are distinct. From Argentina. Released into Florida, the Northeast, and the Midwest, where it has become established. The only parrot to build a stick nest (all others nest in holes), it prefers to build in conifers. Can damage fruit. Raucous noise in nesting colonies.

WHITE-WINGED PARAKEET *Brotogeris versicolurus* 9"
VOICE: Strident, loud screams.

Long tail and bright white and yellow wing patches. Well established (numbers in the hundreds) only in Florida, but escapes have been recorded as far north as Mass.

BUDGERIGAR *Melopsittacus undulatus* to 7"
VOICE: Squawks and cackles. In flocks screams loudly.

Familar to all, the green form is the true color of this Australian native. Declining resident in Florida. Pet escapes seen as far north as New England.

The following species have been seen with some regularity as escapes and could show up at feeders. Native in sites listed.

1. Yellow-headed Parrot. *Amazona oratrix.* Tropical America.
2. Black-hooded Conure (Nanday Conure). *Nandayus nenday.* S. Amer ica.
3. Blossom-headed Parakeet. *Psittacula roseata.* Himalayas.
4. Ring-necked Parakeet. *Psittacula krameri.* India.
5. Yellow-collared Lovebird. *Agapornis personatus.* E. Africa.
6. Hispaniolan Parakeet. *Aratinga chloroptera.* Hispaniola.
7. Green Parakeet. *Aratinga holochlora.* Mexico.
8. Red-crowned Parrot. *Amazona viridigenalis.* Mexico.
9. Orange-fronted Parakeet. *Aratinga canicularis.* Mexico.
10. Orange-chinned Parakeet. *Brotogeris jugularis.* Cen. America.
11. Cockatiel. *Nymphicus hollandicus.* Australia.

CAROLINA PARAKEET
(formerly endemic,
now extinct)

PARROTS (Escapes)

WHITE-
WINGED
PARAKEET

MONK
PARAKEET

BUDGERIGAR
Some individuals
may be blue or yellow

OCCASIONAL ESCAPES

1

2

3

4

5

6

7

8

9

10

11

PIGEONS and DOVES Family Columbidae

Trim to plump, fast-flying birds with small heads. Gregarious. Passenger Pigeon is extinct. FOOD: Seeds, fruits, waste grains, insects.

MOURNING DOVE *Zenaida macroura* 12"
VOICE: A mournful, hollow *coah, cooo, coo, coo.* Often mistaken for an owl calling.
HABITAT: Farms, towns, weed fields, roadsides.
NOTES: Has a long tail with white graduated edge feathers. May breed year-round.

COMMON GROUND-DOVE
Columbina passerina 6½"
VOICE: A soft, repeated *woo-oo, woo-oo.*
HABITAT: Farms, roadsides, fields, wood edges, dunes.
NOTES: Small, sparrowlike; rusty outerwings. Nods head when it walks. Wanders north on occasion.

INCA DOVE *Columbina inca* to 7½"
VOICE: A double *coo-coo.*
Now resident in Texas and Louisiana. Very rare wanderer to north. Roadsides, fields, and waste places. Long tail with white edge. Scaled breast.

WHITE-WINGED DOVE *Zenaida asiatica* 11½"
VOICE: A distinct *who cooks for you.*
Wanders to East Coast in fall and winter. Found in West to e. Texas and in Florida. Square tail, large white wing patches.

WHITE-CROWNED PIGEON *Columba leucocephala* 13"
VOICE: A deep, resonant *coo-ka-croo-coo-coo.*
Extreme s. Florida and Keys species. Large, dark with white to grayish crown. Flocks commute inland from offshore nesting islands for fruit.

RINGED TURTLE-DOVE *Streptopelia risoria* to 12"
VOICE: *Cooo-ka-roo.*
All blond with black neck-ring. Domesticated birds escape on occasion and visit feeders. Formerly established in Florida.

ROCK DOVE (Domestic or Feral Pigeon)
Columba livia to 13"
VOICE: A gurgling *cooo-kura-coo-coo.*
Perhaps the best-known bird in the U.S. Lives everywhere humans live.

PASSENGER PIGEON

extinct

PIGEONS, DOVES
sexes similar

COMMON GROUND-DOVE

INCA DOVE

MOURNING DOVE

WHITE-CROWNED PIGEON

WHITE-WINGED DOVE

plumages variable

ROCK DOVE
(Domestic or Feral Pigeon)

RINGED TURTLE-DOVE

typical or ancestral form

159

CUCKOOS and ALLIES Family Cuculidae

Slender, long-tailed birds. Cuckoos are shy and secretive. **FOOD:** Caterpillars, insects, fruits. Roadrunners eat reptiles.

YELLOW-BILLED CUCKOO
Coccyzus americanus   **to 13"**
VOICE: A rapid, throaty *ka-ka-ka-ka-ka-ka-kow-kow-kow-kowlp-kowlp-kowlp* (slowing at end).
HABITAT: Woodlands, thickets, farms, orchards.
NOTES: Rufous in wings; teardrop undertail spots.

BLACK-BILLED CUCKOO
Coccyzus erythropthalmus   **to 12"**
VOICE: A repetitive, rapid *cucucucu-cucucu*.
HABITAT: Wood edges, groves, thickets.
NOTES: No rufous in wings; bars on undertail.

MANGROVE CUCKOO
Coccyzus minor   **to 12"**
VOICE: A slow, guttural *caow, caow, caow*.
HABITAT: Mangroves and tropical hardwoods of s. Florida.
NOTES: Shy denizen that is difficult to see. Deep buff underbelly and black mask.

SMOOTH-BILLED ANI
Crotophaga ani   **to 12½"**
VOICE: A whining whistle or querulous *que-lick*.
HABITAT: Brushy edges and thickets in s. Florida.
NOTES: Has a long tail and a huge smooth bill. Flight is sloppy. Roost together and preen one another.

GROOVE-BILLED ANI
Crotophaga sulcirostris   **to 13"**
VOICE: A repeated whistled *whee-oo* or *tee-oh*.
HABITAT: Fields, waste places. Confined to the Gulf Coast from Texas to Florida Panhandle.
NOTES: Can be identified by its grooved bill and call.

GREATER ROADRUNNER
Geococcyx californianus   **to 24"**
VOICE: 6–8 descending dovelike *cooos*.
HABITAT: Dry open country.
NOTES: Ground cuckoo that chases reptiles on foot.

CUCKOOS, etc.

sexes similar

YELLOW-BILLED CUCKOO

adult

BLACK-BILLED CUCKOO

MANGROVE CUCKOO

immature

adult

GROOVE-BILLED ANI

SMOOTH-BILLED ANI

GREATER ROADRUNNER

161

GOATSUCKERS (NIGHTJARS) Family Caprimulgidae

Nocturnal birds with ample tails, large eyes, tiny bills, and long rictal (area around mouth) bristles. Goatsuckers rest by day on ground or on limbs and hunt by night. Nighthawks also forage in the daytime. **FOOD:** Insects.

COMMON NIGHTHAWK
Chordeiles minor　　　　　　　　　　　　9½"

VOICE: A nasal *peent* or *pee-ik*.

HABITAT: Openings in woodlands, pine woods, field edges, rooftops in towns and cities.

NOTES: Numbers are decreasing, especially in East. White patches are seen near bend of wing, and long wings extend to tail end. Their wings create a deep roar when the birds dive in display.

WHIP-POOR-WILL
Caprimulgus vociferus　　　　　　　　　9½"

VOICE: A rolling, repeated *whip-poor-weel*.

HABITAT: Open woodlands.

NOTES: More often heard than seen. Numbers fluctuate based on food supply of large insects and moths. When folded, wingtips fall well short of tail end.

CHUCK-WILL'S-WIDOW
Caprimulgus carolinensis　　　　　　　12"

VOICE: A distinct *chuck-will-wid-ow*.

HABITAT: Pine woods, woodlands.

NOTES: Large and buffy with little white in tail.

COMMON POORWILL
Phalaenoptilus nuttallii　　　　　　　to 8"

VOICE: A loud, repeated *poor-will*.

HABITAT: Dry hills and open brush.

NOTES: Barely makes it into plains states from West. Only N. American bird known to hibernate.

GOATSUCKERS

wing of
**LESSER
NIGHTHAWK**
for comparison

♀

**COMMON
NIGHTHAWK**

♂

WHIP-POOR-WILL

♂

**COMMON
NIGHTHAWK**

♂

♂

WHIP-POOR-WILL

♂

**CHUCK-WILL'S-
WIDOW**

♂

**COMMON
POORWILL**

♂

**CHUCK-WILL'S-
WIDOW**

♂

163

HUMMINGBIRDS Family Trochilidae

The smallest birds. Usually iridescent with needlelike bills for sipping nectar. Jewellike throat feathers. Can hover and fly backward. Pugnacious. **FOOD:** Nectar of flowers (red flowers are favored), aphids, and other small insects.

RUBY-THROATED HUMMINGBIRD
Archilochus colubris

3¾" (including bill)

VOICE: In aerial display male flies in pendulum pattern accompanied by hum. Call: squeaks.
HABITAT: Flowers in gardens, wood edges, bogs.
NOTES: The only common eastern hummingbird. Green to gray sides. Male's ruby throat seen only when the sun's angle is right; otherwise throat looks black.

RUFOUS HUMMINGBIRD
Selasphorus rufus   3½" (including bill)
VOICE: High-pitched chips and twitters.
HABITAT: Flowers in gardens, wood edges.
NOTES: Look for this visitor from the West in the late fall or early winter at hummingbird feeders or late flowering salvia (usually females or immatures). Rusty sides and rust in tail. Other western species may wander to East, especially to Gulf Coast states.

KINGFISHERS Family Alcedinidae

Solitary birds with large heads and hefty, pointed bills. American species are fish eaters that hover and then plunge headfirst into water for fish. **FOOD:** Fish.

BELTED KINGFISHER
Ceryle alcyon   13"
VOICE: A loud, dry rattle.
HABITAT: Streams, lakes, bays, coasts. As water freezes in North will move to coast or shift south.
NOTES: A large bird with a crest. Female has additional belly band of rust. Often seen hovering over open water. Excavates a nest hole in banks.

HUMMINGBIRDS

RUBY-THROATED HUMMINGBIRD

♂

♀

Sphinx moth resembles hummingbird

RUFOUS HUMMINGBIRD

♂

♀

KINGFISHERS

hovering

♀

♂

plunging

BELTED KINGFISHER

WOODPECKERS Family Picidae

Woodpeckers are chisel-billed, wood-boring birds. Their feet and stiff tail feathers are designed to anchor them in place on tree trunks. They all excavate holes to nest in. Flight is undulating. Most males have red on head. **FOOD:** Tree-boring insects, berries, acorns, tree sap.

RED-HEADED WOODPECKER
Melanerpes erythrocephalus　　　　**to 9½"**

VOICE: A loud *queer* or *queeah*.

HABITAT: Open woodland, farm country, orchards, shade trees in towns, pond edges with dead trees.

NOTES: The only woodpecker in the East with an all-red head. Has large white wing patches and a white rump. Will establish winter territories defending stored acorns.

PILEATED WOODPECKER
Dryocopus pileatus　　　　**to 19½"**

VOICE: A loud, echoing *kik-kik-kikik-kikkik*. Hammering in deep, well-spaced thuds.

HABITAT: Conifers, mixed woods, hardwoods, ponds and lakes, forest edges.

NOTES: A crow-sized woodpecker with a flaming red crest and a white-striped face. Makes holes that are large rectangles. Southern birds are tamer than those in the North.

IVORY-BILLED WOODPECKER
Campephilus principalis　　　　**to 20"**

VOICE: A loud tooting.

NOTES: Believed to be extinct; there have been no confirmed sightings since the 1950s. Habitat preference was the river bottom country of S.C. to Florida west to eastern Texas.

immature

adult

WOODPECKERS

RED-HEADED
WOODPECKER
sexes similar

♂

♀

under

PILEATED
WOODPECKER

♀

under

♂

IVORY-BILLED
WOODPECKER
extinct

upper

NORTHERN (Yellow-shafted) FLICKER

Colaptes auratus (in part)   **to 14"**

VOICE: A rolling *wick, ka wick, ka wick, ka wick* or squeaky *flick-a, flick-a.*

HABITAT: Open forests, woodlots, farms, towns.

NOTES: A large woodpecker with distinct white rump that flashes when it takes off. Yellow shafting to wing feathers. Male has a black mustache. Fond of ants, so often flushes from ground.

NORTHERN (Red-shafted) FLICKER

Colaptes auratus (in part)

Western color variant. Note salmon shafting to wing feathers. Male has a red mustache. Ranges of these 2 color variants overlap on Great Plains. Very rare in East.

RED-BELLIED WOODPECKER

Melanerpes carolinus   **to 10½"**

VOICE: A nasal *chiv-chiv* or *churr-churr.*

HABITAT: Woodlands, orchards, towns, parks.

NOTES: Laddered back with great deal of red on head. Red on belly is difficult to see. Expanding its range northward.

RED-COCKADED WOODPECKER

Picoides borealis   **to 8½"**

VOICE: A rough, rasping *sripp* or *zhilp.*

HABITAT: Open pine woodlands, especially where pines have red-heart disease.

NOTES: Endangered. Back is zebra-striped; small spot of red on male's head. Very local resident within range.

YELLOW-BELLIED SAPSUCKER

Sphyrapicus varius   **to 9"**

VOICE: A nasal mewing. Drumming—several rapid taps—followed by slow series.

HABITAT: Woodlands, orchards.

NOTES: Large white wing patches and white rump. Drills orderly rows of small holes, often covering trunk of tree, for sap and to attract insect prey.

Yellow-shafted form

Red-shafted

Red-shafted form

♂

♂

Yellow-shafted

♀

NORTHERN FLICKER
Yellow-shafted form

immature

♀

RED-BELLIED WOODPECKER

♂

♂

immature

RED-COCKADED WOODPECKER

♂

YELLOW-BELLIED SAPSUCKER

DOWNY WOODPECKER
Picoides pubescens   **to 6½"**
VOICE: A rapid whinny of descending notes. Also a sharp *pick* (Hairy's is a *peek*).
HABITAT: Forests, woodlands, river groves, shade trees.
NOTES: Small, with small bill. Male has a solid red patch on back of head. Black flecks in outer tail feathers. Common birdfeeder species.

HAIRY WOODPECKER
Picoides villosus   **to 9½"**
VOICE: A fast, dry rattle. Note is a sharp *peek*.
HABITAT: Forests, woodlands, river groves, parks.
NOTES: A mid-sized woodpecker with a larger bill than the Downy. Male's red patch has a black line through it. Outer tail feathers are pure white. Numbers are dropping in many areas in the East.

THREE-TOED WOODPECKER
Picoides tridactylus   **to 9½"**
VOICE: A sharp *kick* or *tick*.
HABITAT: Conifer forests and bogs.
NOTES: Note yellow cap (male), laddered back. Only 3 toes. Chips bark scales from trees rather than digging holes. Rare wanderer south to dash line.

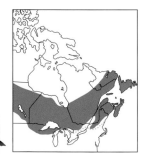

BLACK-BACKED WOODPECKER
Picoides arcticus   **to 10"**
VOICE: A sharp *chuck*.
HABITAT: Conifer forests, bogs.
NOTES: Has a yellow cap (male), 3 toes, and a black back. Lacks white in wings. Leaves distinctive marks on trunks of trees by chipping off bark. More common than the Three-toed Woodpecker. Rare wanderer south to dash line.

WOODPECKERS

DOWNY WOODPECKER

♀
♂

HAIRY WOODPECKER

♀
♂
♂ southern form

THREE-TOED WOODPECKER

♀
♂

BLACK-BACKED WOODPECKER

♀
♂

171

TYRANT FLYCATCHERS Family Tyrannidae

Most flycatchers sit upright and sally forth to snap up insects. Heavy flattened bill with bristles at base. **FOOD:** Insects—adult and larvae.

SCISSOR-TAILED FLYCATCHER

Tyrannus forficatus   **to 15"**

VOICE: Harsh *keck* or bickering *ka-leep.*

HABITAT: Open country, plains, ranches, wires.

NOTES: Young have shorter tails but also show pink sides. Rare straggler to North and East.

EASTERN KINGBIRD *Tyrannus tyrannus*   **8"**

VOICE: Rapid *kit-kit-kitter*, nasal *dzeep.*

HABITAT: Wood edges, river groves, orchards, farmlands, roadsides.

NOTES: White-tipped tail. Commonly darts out to snatch insects. Pugnacious.

WESTERN KINGBIRD *Tyrannus verticalis* **8"**

VOICE: Sharp *whit* or *whit-ker-whit.*

HABITAT: Open country with scattered trees.

NOTES: Has white outer tail feathers. Perches on fences and powerlines along roadsides. Wanders east in fall.

GRAY KINGBIRD

Tyrannus dominicensis   **to 9"**

VOICE: A rolling *pi-teer-rry.*

HABITAT: Mangroves, wood edges, on wires.

NOTES: Huge bill, deep forked tail. Florida specialty that has wandered as far north as New England.

GREAT CRESTED FLYCATCHER

Myiarchus crinitus   **to 9"**

VOICE: A loud *wheeep;* rolling *prrrrrreet!*

HABITAT: Woodlands, groves, suburban woodlots.

NOTES: Rust in tail and wings. Deep gray throat. Nests in holes.

ASH-THROATED FLYCATCHER

Myiarchus cinerascens   **8"**

Western. Rare in East in fall. Like small Great Crested, but grayer back, with white throat and pale yellow belly.

172

LARGE FLYCATCHERS

sexes similar

EASTERN KINGBIRD

SCISSOR-TAILED FLYCATCHER

ASH-THROATED FLYCATCHER

WESTERN KINGBIRD

GREAT CRESTED FLYCATCHER

GRAY KINGBIRD

173

EASTERN PHOEBE *Sayornis phoebe* to 7"

VOICE: Song is a well-enunciated, rising *phoe-be*; call is a *chip*.

HABITAT: Bridges, streamsides, wood edges, towns.

NOTES: Wags tail downward. Immatures may show pale wingbars and pale lemon underbelly.

EASTERN WOOD-PEWEE

Contopus virens to 6½"

VOICE: A drawn out, rising *peeeee-a-wee*, and a descending *pee-ur*.

HABITAT: Woodlands, groves.

NOTES: Pale wingbars, pale lower mandible. Does *not* flick tail.

LEAST FLYCATCHER *Empidonax minimus*

Placed here for comparison. Note eye-ring.

OLIVE-SIDED FLYCATCHER

Contopus cooperi to 8"

VOICE: Note is a *pip-pip-pip*; call is a distinct *quick-three-beers* with middle note highest.

HABITAT: Conifer forests, bogs, slash areas.

NOTES: Dark sides to chest. White tufts not easy to see. Sits atop dead tree tops and snags.

SAY'S PHOEBE *Sayornis saya* to 8"

Rare wanderer from the West. Note black tail and orange underbelly.

VERMILION FLYCATCHER

Pyrocephalus rubinus 6"

Small. Adult males show brilliant colors; immatures and females show streaking and pink underbelly. Few wander to Gulf Coast from Louisiana to Florida each winter.

FLYCATCHERS

sexes similar except in
Vermilion Flycatcher

**EASTERN
PHOEBE**

**EASTERN
WOOD-PEWEE**

**LEAST
FLYCATCHER**

Least Flycatcher is a
typical Empidonax.
See others on p. 177

**OLIVE-SIDED
FLYCATCHER**

**SAY'S
PHOEBE**

**VERMILION
FLYCATCHER**

♀

**VERMILION
FLYCATCHER**

♂

THE EMPIDONAX COMPLEX Genus *Empidonax*

These 5 small flycatchers all have a light eye-ring and 2 whitish wing-bars. During nesting season they can be identified by their song. They seldom sing during fall migration, however, making identification more difficult. **FOOD:** Insects and larvae.

ACADIAN FLYCATCHER

Empidonax virescens   **to 4½"**

VOICE: An explosive *pit-see* (upward inflection).
HABITAT: Evergreen and deciduous forests, ravines, swampy lowlands, beech woods.
NOTES: Plumage greenish above. Range spreading northeast.

YELLOW-BELLIED FLYCATCHER

Empidonax flaviventris   **5½"**

VOICE: A light *per-wee* or *chu-wee*, also a sneezy *chew*.
HABITAT: Woodland edges in migration. Breeds in northern muskeg, bogs.
NOTES: Yellowish underparts and eye-ring. Typically is a late-spring migrant.

LEAST FLYCATCHER

Empidonax minimus   **5½"**

VOICE: Song a sharp, emphatic, *che-bek*; call a *whit*.
HABITAT: Woods, river groves, orchards, shade trees.
NOTES: Gray with white throat.

WILLOW FLYCATCHER

Empidonax traillii   **5½"**

VOICE: Song a wheezy, quick *fitz-bew*; call a soft *whit*.
HABITAT: Field and marsh edges, bushes, willow thickets.
NOTES: Almost identical to Alder Flycatcher. Best distinguished by its call.

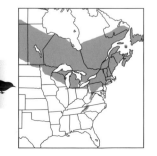

ALDER FLYCATCHER

Empidonax alnorum   **5½"**

VOICE: Song a *fee-bee-o* or *pi-pit*; call *peep* or *pip*.
HABITAT: Willows, alders, brushy swamps.
NOTES: Best distinguished from Willow by call. In migration, when silent, nearly impossible to separate the two.

EMPIDONAX FLYCATCHERS

pit-see!

All 5 Empidonax flycatchers have eye-rings and wingbars. Identify by habitat and voice.

sexes similar

chu-wee

deciduous woods, esp. beech trees; wooded swamps; s. and cen. U.S.

conifer woods, bogs; Canada, n. edge of U.S.

ACADIAN
greener than Least, Alder, or Willow

che-BEK or chebek

YELLOW-BELLIED
breast washed with yellow

fitz-bew

farms, orchards, groves, open woods; n. U.S. and Canada

LEAST
grayest of the group

fee-bee'-o

wet and dry thickets, brushy pastures, old orchards, willows; n. and cen. U.S.

WILLOW

alder swamps, wet thickets, usually near water; n. U.S., Canada

ALDER

177

LARKS Family Alaudidae

Larks are ground birds that often sing in flight. They are brown and have a bold face pattern. **FOOD:** Mainly seeds, insects.

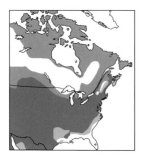

HORNED LARK *Eremophila alpestris* **to 8"**
VOICE: A tinkling that is prolonged and high-pitched. Sometimes delivered in the air. Note is *tsee-titi*.
HABITAT: Open fields, prairies, golf courses, airports, shores, and tundra.
NOTES: Walks, often pressing close to the ground. In flight it reveals a black tail and folds wings after each beat. A local breeder in the East.

PIPITS Family Motacillidae

Streaked ground birds with white outer tail feathers. Pipits walk rather than hop, bobbing the head and tail constantly. **FOOD:** Insects, seeds.

AMERICAN PIPIT *Anthus rubescens* **to 7"**
VOICE: A *pip-pit* or *jee-eet*. Aerial display and *chwee-chwee-chwee* at nest site.
HABITAT: Tundra, alpine slopes of mountains. Plowed fields and shorelines in migration.
NOTES: More often seen flying overhead calling rather than on the ground. Often in flocks in fall and winter. Bobs its tail as it walks.

SPRAGUE'S PIPIT *Anthus spragueii* **to 6½"**
VOICE: A series of very high-pitched descending notes delivered high in the air: *tink-a-ling-tink-a-ling*.
HABITAT: Plains, short-grass prairies.
NOTES: Pale legs, plain face, buffy breast, and patterned back. Very sparrow-like. A skulker that can be difficult to flush. In flight will roller coast, then plunge straight down to the ground.

LARKS, PIPITS
sexes similar

immature

prairie
race

overhead

northern
race

**HORNED
LARK**

topside

overhead

tail-wagging

summer

winter

**AMERICAN
PIPIT**

overhead

towering
flight

SPRAGUE'S PIPIT

SWALLOWS Family Hirundinidae

Swallows have a slim, streamlined form and graceful flight. They have pointed wings and short bills. **FOOD:** Flying insects are caught with wide mouth open during flight. Tree Swallows also eat bayberries.

PURPLE MARTIN *Progne subis*   to 8½"
VOICE: A throaty, rich, bubbling *tchew-wew.*
HABITAT: Towns, farms, open country often near water. Attracted to martin boxes and special nesting gourds. Colonial and local.
NOTES: Largest N. American swallow. Glides in circle patterns alternating flapping and gliding.

CLIFF SWALLOW
Petrochelidon pyrrhonota   to 6"
VOICE: A *zayrp* or low *churr.* Sharp *keer* is an alarm note at nest site.
HABITAT: Open land, farms, cliffs, river bluffs.
NOTES: Nests colonially on cliffs, buildings, and under bridges. Nesting localized in many areas. Colonies may disappear for years for no apparent reason.

BARN SWALLOW *Hirundo rustica*   7¾"
VOICE: A soft *vit* or *kvik-kvik, vit-vit.* Anxious notes and twittering around nest sites.
HABITAT: Open area, farms, garages, fields, marshes. Roosts on wires.
NOTES: The only swallow with a long, deeply forked tail. Often flies very close to the ground catching insects. Adapts well to areas populated by humans.

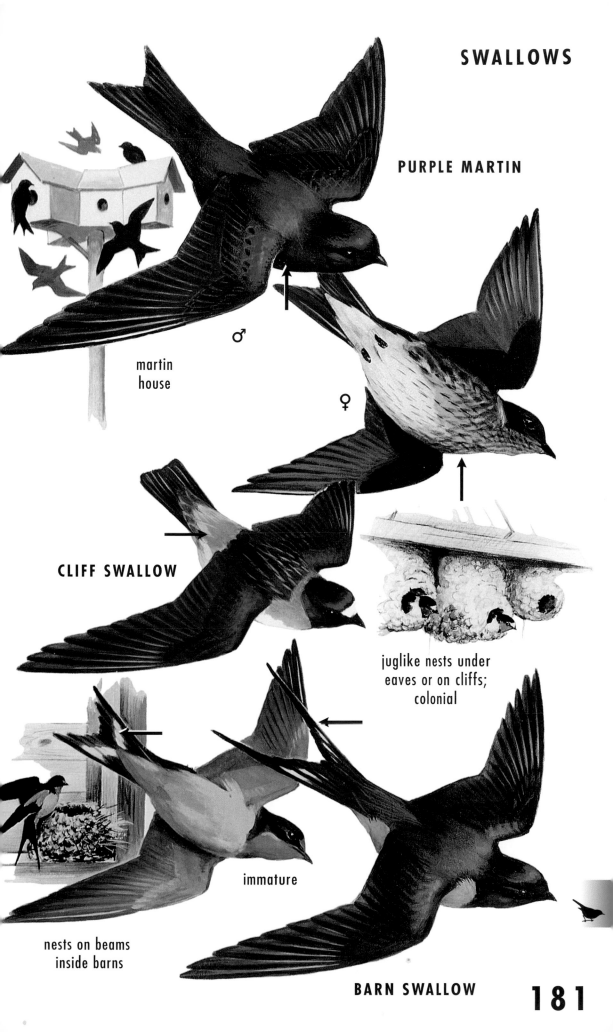

SWALLOWS

PURPLE MARTIN

martin
house

♂

♀

CLIFF SWALLOW

juglike nests under
eaves or on cliffs;
colonial

immature

nests on beams
inside barns

BARN SWALLOW

TREE SWALLOW *Tachycineta bicolor* **to 6"**
VOICE: A liquid *weet-trit-weet*. Call: *chee-vit*.
HABITAT: Open country near water, marshes, meadows, stream edges. Roosts on wires and in marsh grasses, often in massive numbers.
NOTES: White throat, no breast band. Takes to boxes put out for them and bluebirds.

NORTHERN ROUGH-WINGED SWALLOW
Stelgidopteryx serripennis **to 5¾"**
VOICE: Burry *trrit*.
HABITAT: Near streams, ponds, lake edges, open areas.
NOTES: Brown above with a dusky throat. Nests in openings in walls and rock crevices.

BANK SWALLOW *Riparia riparia* **to 5½"**
VOICE: A dry rattle: *trr-tri-tri*.
HABITAT: Sand banks, fields, streams, open areas.
NOTES: Bold breast band. Nests in holes excavated in sand banks. Large colonies are local and sporadic.

SWIFTS Family Apodidae

Swifts look like cigars with wings; they have stiff wing movements and erratic flight with rapid wingbeats. Their wings are often bowed in a crescent shape. **FOOD:** Flying insects.

CHIMNEY SWIFT *Chaetura pelagica* **to 5½"**
VOICE: A loud, rapid ticking and twittering.
HABITAT: Skies over cities, towns, and open country.
NOTES: Has adapted to living in manmade structures but still nests in trees in some areas.

VAUX'S SWIFT *Chaetura vauxi* **to 4½"**
A small swift shown for comparison. Note pale throat. A western species that is very rare in East. May be overlooked, especially along Gulf Coast.

TREE SWALLOW

adult

sexes similar

nests in tree holes or bird boxes

TREE SWALLOW

immature

Bank Swallow colony

BANK SWALLOW

NORTHERN ROUGH-WINGED SWALLOW

VAUX'S SWIFT

CHIMNEY SWIFT

roosting swifts

Purple Martin

♂

Barn

Bank

Northern Rough-winged

Tree

Cliff

swallows on a wire

CROWS, JAYS, etc. Family Corvidae

Large- to medium-sized birds. Their strong bills are covered with bristles at the base. Sexes look alike. FOOD: Nearly anything edible.

FISH CROW *Corvus ossifragus* 16" to 20"
VOICE: A nasal, *qua* or *qua-ha*.
HABITAT: Coasts, rivers, tidewaters, and fields.
NOTES: Smaller and trimmer than the American Crow. Fish Crow is best distinguished by its voice.

AMERICAN CROW
Corvus brachyrhynchos 17" to 21"
VOICE: The classic *caw*, plus a wide variety of other guttural sounds. Has its own "language."
HABITAT: Woodlands, farmlands, parks, shores. Has adapted to areas populated by humans.
NOTES: One of the best-known wild birds, the American Crow forms winter roosts of thousands.

CHIHUAHUAN RAVEN
Corvus cryptoleucus to 21"
VOICE: A high, hoarse *kraahk*.
HABITAT: Arid plains, dry flatlands.
NOTES: Barely reaches area covered by this book in Oklahoma, Texas, and Kansas.

COMMON RAVEN *Corvus corax* to 27"
VOICE: A croaking *cr-r-ruck* and a clear *tok*.
HABITAT: Mature northern and mountain forests, coastal cliffs, tundra, mountain ridges.
NOTES: Wedge-shaped tail. Expanding its range.

Fish Crow

FISH CROW

American Crow

CHIHUAHUAN RAVEN

AMERICAN CROW

Common Raven

COMMON RAVEN

185

BLUE JAY *Cyanocitta cristata* **to 12½"**

VOICE: A harsh, slurred *jay*, a musical *queedle*.

HABITAT: Oak and pine woods, suburban gardens, parks, towns; has adapted to humans.

NOTES: Large, colorful, and well known, the Blue Jay moves in large groups in the fall. Mimics hawk calls and loves to feed on acorns it stores for winter.

FLORIDA SCRUB-JAY

Aphelocoma coerulescens **to 12"**

VOICE: A rasping *kwesh-kwesh*, also *zhrink*.

HABITAT: Scrub oak and pine woods.

NOTES: Has now been separated as a full species from 2 other western species.

GRAY JAY

Perisoreus canadensis **to 13"**

VOICE: A soft *whee-ah*. Also makes gurgled notes and even chatters harshly.

HABITAT: Boreal fir and spruce forests.

NOTES: Looks like a huge chickadee! Often very tame. Visits campsites for handouts.

BLACK-BILLED MAGPIE *Pica pica*
to 22" (including long green tail)

VOICE: A harsh, rapid *queg, queg, queg, queg,* or nasal *maaag* or *aag-aaag*.

HABITAT: Rangeland, open country, woodland edges, streamsides.

NOTES: Rare to the East, the Black-billed Magpie makes a huge ball-like nest of sticks.

JAYS,
MAGPIES

sexes alike

BLUE JAY

FLORIDA
SCRUB-JAY

GRAY JAY

adult

GRAY JAY
juvenile

BLACK-BILLED
MAGPIE

TITMICE Family Paridae

Small, plump birds. Often acrobatic while feeding. Very often travel in groups in mixed-species flocks with kinglets and warblers. **FOOD:** Insects, seeds, berries. A standard group at feeders, titmice prefer sunflower seeds.

BLACK-CAPPED CHICKADEE
Poecile atricapillus   **to 5½"**
VOICE: A distinct *chic-a-dee-dee-dee*. Song is a clear, whistled *fee-bee*, with the first note higher.
HABITAT: Woods, thickets, parks, feeders.
NOTES: These trusting birds often forage in flocks. In fall and winter they often mix with other woodland species. Periodic winter influxes southward.

CAROLINA CHICKADEE
Poecile carolinensis   **4½"**
VOICE: A high, rapid *tish-ta-dee-dee*, delivered faster than the notes of the Black-capped. Call is 4-parted *fee-bee-fee-bay*.
HABITAT: Woodlands, pinewoods, parks; visits feeders. Range overlaps with Black-capped at northern edge. Can learn calls from other species.
NOTES: Has a small bib; less white in wing. Hybrids have occurred.

BOREAL CHICKADEE
Poecile hudsonicus   **to 5½"**
VOICE: A slow, drawn out, raspy *chick-che-day-day*.
HABITAT: Conifer forests of the North.
NOTES: Often tame, the Boreal has a brown cap and extensive pinkish brown sides. Sporadically wanders south in winter.

TUFTED TITMOUSE *Baeolophus bicolor*   **6"**
VOICE: A clearly whistled *peter-peter-peter* or *here-here-here*. Also delivers a variety of whistles and wheezy plaintive notes.
HABITAT: Woodlands, parks, feeders.
NOTES: A stalwart at the bird feeder, the Tufted Titmouse has a perky crest and black button eyes. Its population is expanding in many areas.

BLACK-CAPPED
CHICKADEE

CAROLINA
CHICKADEE

BOREAL
CHICKADEE

immature

adult

TUFTED
TITMOUSE

NUTHATCHES Family Sittidae

Small, stout tree hunters with strong bills, nuthatches often climb headfirst down tree trunks or forage upside down on limbs. **FOOD:** Insect eggs, seeds, nuts. At feeders, nuthatches are attracted to suet and sunflower seeds.

WHITE-BREASTED NUTHATCH
Sitta carolinensis **to 6"**
VOICE: A series of rapid low, nasal *yank-yank-yank*s.
HABITAT: Forests, woodlots, parks, feeders.
NOTES: Male has a black cap. Inquisitive.

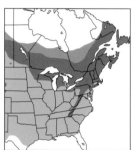

RED-BREASTED NUTHATCH
Sitta canadensis **to 4½"**
VOICE: A high, fast, repeated *ank-ank-ank*.
HABITAT: Conifer and mixed woods. Large influxes southward during some winters.
NOTES: Explores tree limbs more than trunks. Flits wings when excited.

BROWN-HEADED NUTHATCH
Sitta pusilla **to 4½"**
VOICE: A high rapid *kit-kit-kit* and a piping *ki-day-ki-day*. Groups twitter and chatter.
HABITAT: Open pine woods. Often in small bands.
NOTES: Often heard from high in pines before seen.

CREEPERS Family Certhiidae

Creepers are small, slim, and stiff-tailed. Their plumage allows them to blend well with tree bark. Curved bill aids in probing. **FOOD:** Insect eggs, insects.

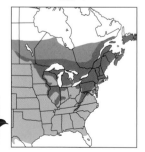

BROWN CREEPER *Certhia americana* **5"**
VOICE: A single, high *seee*. Song is a twittering jumble of musical notes.
HABITAT: Woodlands, swamps, parks.
NOTES: Forages by creeping up trees and then dropping to base of next tree and spiraling up. Hides nest behind loose flap of bark. Scattered nesting records south of mapped range.

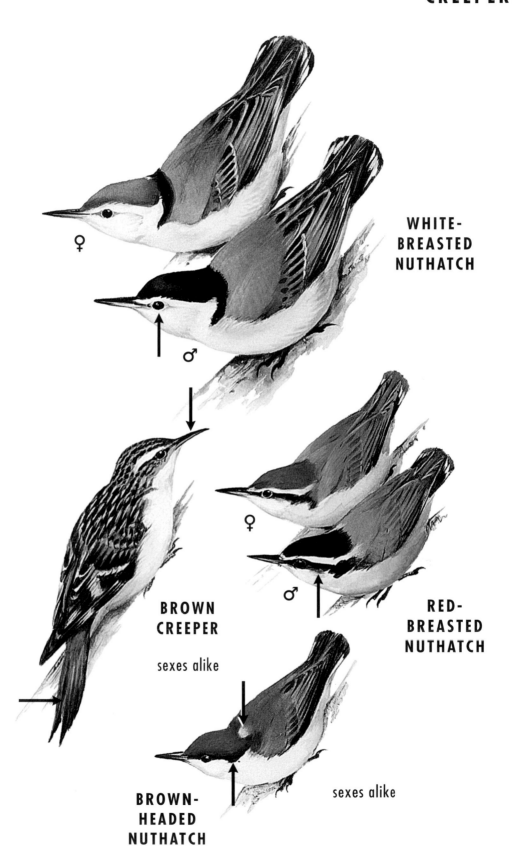

♀

**WHITE-
BREASTED
NUTHATCH**

♂

**BROWN
CREEPER**

sexes alike

♀

♂

**RED-
BREASTED
NUTHATCH**

sexes alike

**BROWN-
HEADED
NUTHATCH**

WRENS Family Troglodytidae

Small, energetic, furtive brown birds. Slightly curved bill; tail is held cocked up. FOOD: Insects and spiders. Will take suet at feeders.

HOUSE WREN *Troglodytes aedon*　　　to 5"
VOICE: A stuttering musical song that rises in the middle and falls at the end.
HABITAT: Wood edges, fields, towns, gardens.
NOTES: Can dominate all bird boxes in a yard.

WINTER WREN *Troglodytes troglodytes*　　4"
VOICE: A beautiful long series of tinkling notes.
HABITAT: Conifer forests, underbrush, brush piles.
NOTES: Tiny skulker of shadowed areas.

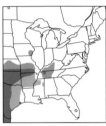

BEWICK'S WREN
Thryomanes bewickii　　　to 5⅕"
VOICE: A series of notes dropping slightly at the end.
HABITAT: Thickets, underbrush, gardens.
NOTES: White edge to tail. Overall decline.

CAROLINA WREN
Thryothorus ludovicianus　　　5⁷⁄₁₀"
VOICE: A clear *tea-kettle, tea-kettle, tea-kettle*.
HABITAT: Brush, tangles, parks, towns, gardens.
NOTES: Populations are reduced by severe winters. Sporadic north to dash line.

MARSH WREN *Cistothorus palustris*　　5"
VOICE: A reedy, gurgling, pulsing series of notes.
HABITAT: Cattail, bulrush, and salt marshes.
NOTES: Makes many domed dummy nests.

SEDGE WREN *Cistothorus platensis*　　4½"
VOICE: A dry staccato chattering *chap-chap-churr*.
HABITAT: Wet swales, grassy sedge meadows, marshes.
NOTES: Very selective of breeding sites; irregular south and east to dash line.

ROCK WREN *Salpinctes obsoletus*　　to 6½"
VOICE: A downward trill.
HABITAT: Rocky slopes and canyons.
NOTES: A rare visitor from the West. Breeds in w. Dakotas to Oklahoma.

HOUSE WREN

WINTER WREN

BEWICK'S WREN

CAROLINA WREN

MARSH WREN

SEDGE WREN

ROCK WREN

GNATCATCHERS and KINGLETS Family Sylviidae

Tiny, active birds. Some move in foraging flocks. Gnatcatchers have long mobile tails they switch back and forth. **FOOD:** Insects, spiders.

RUBY-CROWNED KINGLET
Regulus calendula **4"**
VOICE: Call is a *ji-dit*. Song is a rollicking and loud *tee tee tee, tew tew tew, ti-dadee, ti-dadee*.
HABITAT: Breeds in conifers. Rest of year in all types of woodlands and brushy borders.
NOTES: Twitches wings. Crown patch often difficult to see.

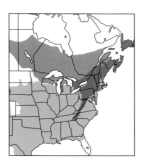

GOLDEN-CROWNED KINGLET
Regulus satrapa **3½"**
VOICE: Call is a high-pitched *see-see-see*; song is a series of thin notes that rise, then drop off.
HABITAT: Breeds in conifers; spends the rest of the year in other woodlands.
NOTES: Twitches wings, though not as frequently as Ruby-crowned. Moves about in flocks.

BLUE-GRAY GNATCATCHER
Polioptila caerulea **4½"**
VOICE: Call is a thin, high *spit-chee*; song is a thin, squeaky, wheezy series of notes.
HABITAT: Open woods, oaks, evergreens.
NOTES: Expanding its range northward. Very active while foraging: it flutters and runs out on limbs, flicking its tail side to side.

BULBULS Family Pycnonotidae

This bulbul was introduced from se. Asia and have survived in s. Florida over the past 40 years.

RED-WHISKERED BULBUL
Pycnonotus jocosus **to 7"**
Sleek-looking with distinct cheek patch. Established locally in Kendall area of s. Miami. Lives in gardens and ornamental plantings. Adapted to live in cities.

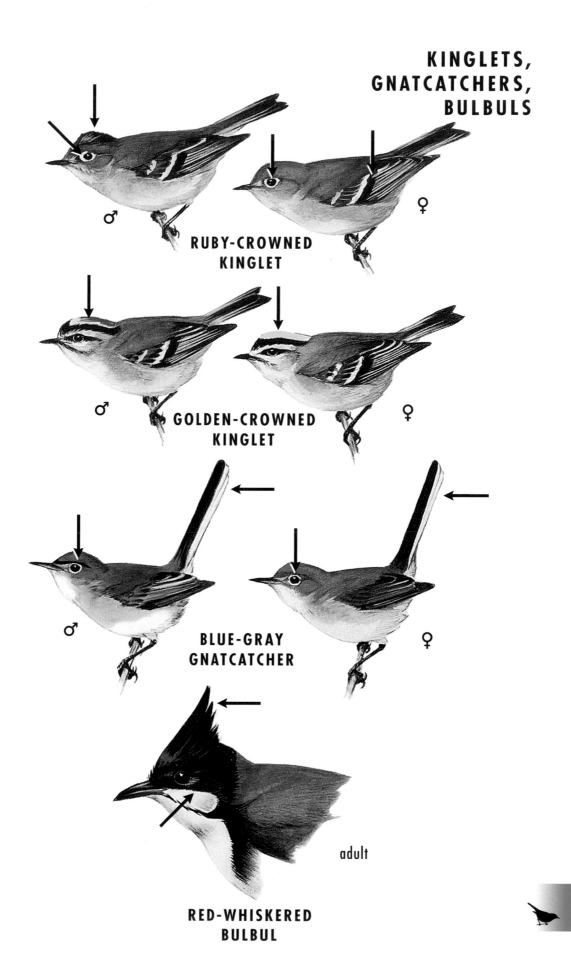

♂

♀

RUBY-CROWNED KINGLET

♂

♀

GOLDEN-CROWNED KINGLET

♂

♀

BLUE-GRAY GNATCATCHER

adult

RED-WHISKERED BULBUL

MOCKINGBIRDS and THRASHERS Family Mimidae

Excellent songsters. Mockingbird mimics other species and may sing at night. **FOOD:** Fruits and insects.

BROWN THRASHER *Toxostoma rufum* 11½"
VOICE: A succession of clear notes and raspy phrases sung in couplets.
HABITAT: Thickets, brush, thorn scrub.
NOTES: Runs on ground in pursuit of insects.

GRAY CATBIRD *Dumetella carolinensis* 9"
VOICE: Catlike mewing, grating *check-check*. Song is a series of disjunct phrases and notes.
HABITAT: Undergrowth, brush, scrub, gardens.
NOTES: Abundant thicket species. Flicks tail as it peers from thickets.

NORTHERN MOCKINGBIRD
Mimus polyglottos **to 11"**
VOICE: A jumbled series of notes and phrases, often mimicking other species. Unmated males sing through the night.
HABITAT: Towns, field edges, parks, overgrown farm fields. Readily suburbanized.
NOTES: Has spread north, rarely and irregularly to dash line.

THRUSHES Family Turdidae

Large-eyed, slender-billed, round-bodied songbirds. All young have spotted breasts. Excellent singers. **FOOD:** Insects, worms, snails, fruit.

TOWNSEND'S SOLITAIRE
Myadestes townsendi 8"
Placed here for comparison to mockingbird. A very rare visitor from the West. Distinguished by its eye-ring and buff wing patches. Winter has been the best time for recording this western straggler.

MOCKINGBIRDS, THRASHERS

sexes alike

BROWN THRASHER

GRAY CATBIRD

NORTHERN MOCKINGBIRD

flight pattern

wing flashing

shrike for comparison

TOWNSEND'S SOLITAIRE
(a thrush)

EASTERN BLUEBIRD *Sialia sialis* 7"

VOICE: A soft, musical churring and *chur-wi*.
HABITAT: Open areas with scattered trees, field edges, orchards. Takes well to nesting boxes.
NOTES: This well-known species has benefited from nesting box programs that have led to its recovery in many areas.

MOUNTAIN BLUEBIRD
Sialia currucoides 7"

This rare winter visitor from the West has a slightly longer bill and is turquoise blue or grayish with blue rump.

AMERICAN ROBIN
Turdus migratorius to 11"

VOICE: A variety of clear, caroling phrases that rise and fall, especially at dawn and dusk. Call: a soft *tup tup*.
HABITAT: Cities, towns, parks, farmlands, field edges, woods. Adapts well to humans.
NOTES: Our most familiar songbird. Builds mud-lined nests in tree crotches on flat surfaces.

VARIED THRUSH *Ixoreus naevius* to 10"

Shown here to compare to American Robin, as this rare visitor from the West appears with some regularity at eastern bird feeders in winter. Has orange wingbars and a breast band.

NORTHERN WHEATEAR
Oenanthe oenanthe 6"

VOICE: A *chack, chack* or *chak-wheet-eeeer*.
HABITAT: Breeds on rocky tundra.
NOTES: Autumn stray to East Coast. Visits coastal areas, dunes, and open rocky farmlands. Winters in Old World.

THRUSHES

♂

♀

juvenile

**EASTERN
BLUEBIRD**

♂

**MOUNTAIN
BLUEBIRD**

♂

♀

juvenile

**AMERICAN
ROBIN**

♂

**VARIED
THRUSH**

summer

winter

**NORTHERN
WHEATEAR**

GRAY-CHEEKED THRUSH
Catharus minimus   **to 8"**
VOICE: A mellow *whee-wheeoo-titi-whee*. Call note is a sharp *quee-ahh*.
HABITAT: Boreal forests, tundra scrub, woods.
NOTES: Bicknell's Thrush (*C. bicknelli*, in red on map), a breeder of mountaintop forests from New York to Nova Scotia, is now regarded as a full species.

SWAINSON'S THRUSH
Catharus ustulatus   **7"**
VOICE: Flutelike phrases that slide upward: *too-lee-toosee-to-eeeeee*. Call note a *whit*.
HABITAT: Spruce forests, woodlands.
NOTES: Distinct buffy eye-ring.

HERMIT THRUSH   *Catharus guttatus*   **7"**
VOICE: 3 to 4 ethereal, flutelike phrases produced at different pitches. Call note: *chuck*.
HABITAT: Conifers, mixed woods, swamps.
NOTES: Cocks tail, then slowly lowers it. Often flicks wings when perched. The spot-breasted thrush of winter.

VEERY   *Catharus fuscescens*   **7½"**
VOICE: A liquid, down-wheeling *vee-ur, vee-u-vee-eee*. Call note is a distinct *veeer*.
HABITAT: Damp, deciduous woods.
NOTES: Least spotted of the thrushes.

WOOD THRUSH   *Hylocichla mustelina*   **8"**
VOICE: Varied flutelike notes: *ee-o-lay-ee-o-lee*. Call note: *pip-pip-pip*.
HABITAT: Deciduous woods.
NOTES: The Wood Thrush, which has bold chest marks and a rusty head, is a distinctive voice of the forest understory.

BROWN THRUSHES

sexes alike

GRAY-CHEEKED THRUSH

SWAINSON'S THRUSH

HERMIT THRUSH

VEERY

WOOD THRUSH

juvenile

SHRIKES Family Laniidae

Songbirds with hook-tipped bills and bird-of-prey behavior. Perch watchfully on treetops and wires and drop on prey, then impale the prey on thorns, barbed wire. **FOOD:** Insects, lizards, mice, small birds.

NORTHERN SHRIKE
Lanius excubitor **to 10"**
VOICE: A succession of harsh, jumbled notes.
HABITAT: Open country with lookout posts.
NOTES: Light vermiculations (bars) on breast. Moves south sporadically in winter to dash line.

LOGGERHEAD SHRIKE
Lanius ludovicianus **9"**
VOICE: Repeated harsh, deliberate phrases *queedle-queedle* or *tsurp-tsurp*.
HABITAT: Power lines and open country with lookout posts or barbed-wire fences.
NOTES: Fast disappearing from many areas, especially in Northeast. Reason for decline is unknown.

WAXWINGS Family Bombycillidae

Sleek, crested, gregarious birds with red waxy tips to their secondaries. **FOOD:** Berries most of year. Waxwings "hawk" out over water and open areas for insects in summer and fall.

BOHEMIAN WAXWING
Bombycilla garrulus **8"**
VOICE: A high-pitched, trembling *zreee*.
HABITAT: Boreal forests and muskeg.
NOTES: Sporadically invades in flocks from the Northwest between dash lines. Starlinglike shape in flight.

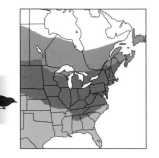

CEDAR WAXWING
Bombycilla cedrorum **7"**
VOICE: A high, ethereal, lisped *tssseeeeee*.
HABITAT: Open woodlands, river edges, orchards. Irregular movement throughout the year.
NOTES: Tail tip may be orange in immature. Move about in large flocks in fall and winter.

SHRIKES
sexes similar

immature

NORTHERN SHRIKE

LOGGERHEAD SHRIKE

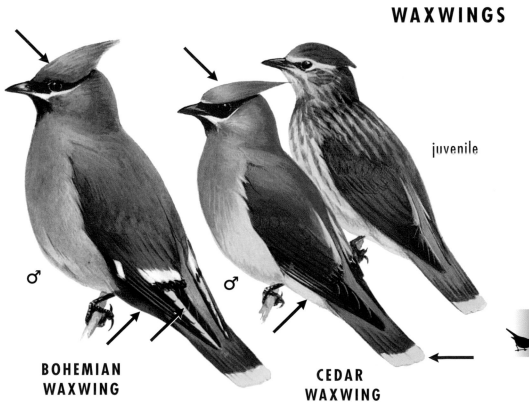

WAXWINGS

juvenile

♂

♂

**BOHEMIAN
WAXWING**

**CEDAR
WAXWING**

VIREOS Family Vireonidae

Small olive or grayish birds the size of warblers. More deliberate in movements. Can be divided into those species with wingbars and those that lack wingbars. **FOOD:** Mostly insects; berries.

RED-EYED VIREO *Vireo olivaceus*   6"
VOICE: Short, abrupt phrases separated by deliberate pauses. Repeated up to 40 times per minute.
HABITAT: Woodlands, shade trees, parks.
NOTES: Their song is the song of eastern summer woodlands. The nest is a hanging cup of plant fibers.

BLACK-WHISKERED VIREO
Vireo altiloquus   5"
VOICE: Similar to that of Red-eyed Vireo.
HABITAT: Mangroves and tropical hardwoods of Florida.
NOTES: Rare visitor along Gulf Coast west to Louisiana.

WARBLING VIREO *Vireo gilvus*   5"
VOICE: A series of clear warbled notes.
HABITAT: Mixed deciduous woodlands, shade trees along waterways, and lake edges.
NOTES: The most drab of the vireos. Male often sings while sitting on nest. Local.

PHILADELPHIA VIREO
Vireo philadelphicus   4$\frac{7}{10}$"
VOICE: Sounds like a high-pitched, slow Red-eyed Vireo.
HABITAT: Poplars, willows, alders in wet areas.
NOTES: Yellow-tinted underparts, especially throat and center of breast. Dark line through lores. Seen mainly as a migrant in U.S.

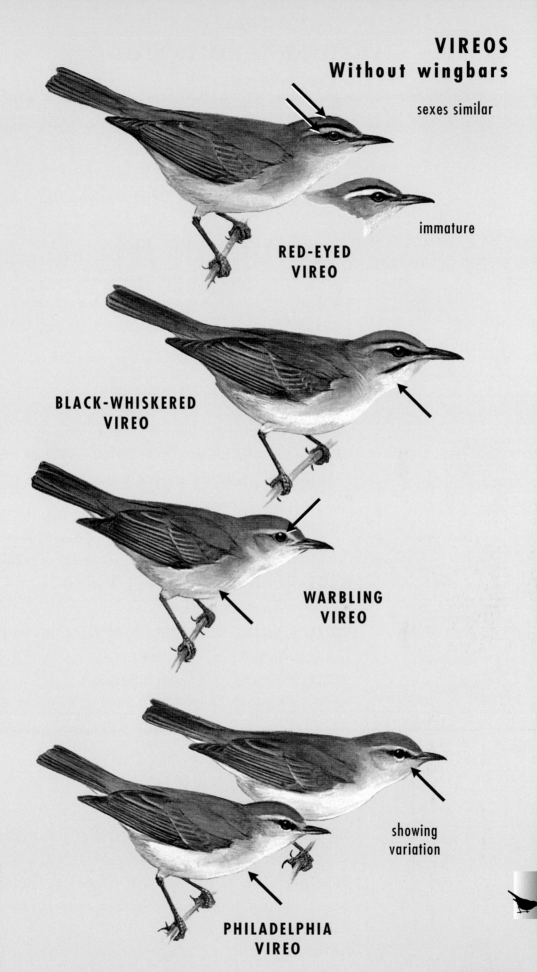

immature

RED-EYED
VIREO

BLACK-WHISKERED
VIREO

WARBLING
VIREO

showing
variation

PHILADELPHIA
VIREO

YELLOW-THROATED VIREO
Vireo flavifrons   5"
VOICE: Similar to that of the Red-eyed Vireo but slower and burry and in a swinging back and forth pattern. Key phrase sounds like *three-eight*.
HABITAT: Shade trees, woodlands, river edges.
NOTES: A treetop species. Inquisitive. Sings into fall.

WHITE-EYED VIREO *Vireo griseus*   5"
VOICE: Distinct *chick-a-pweer-chick*.
HABITAT: Thickets, edges, overgrown fields.
NOTES: Absent at higher elevations. Young have dark eyes. May be difficult to see.

BELL'S VIREO *Vireo bellii*   5"
VOICE: A series of question and answer phrases: *cheedle-cheedle-chee? cheedle-cheedle-chew.* The first phrase rises; the second falls.
HABITAT: Willows, streamsides, thickets.
NOTES: Expanding its range eastward.

BLUE-HEADED (Solitary) VIREO
Vireo solitarius   6"
VOICE: Like that of the Red-eyed Vireo but with long hesitations between phrases.
HABITAT: Mixed conifer-deciduous woods.
NOTES: Blue-headed, Plumbeous, and Cassin's forms recently split into species.

BLACK-CAPPED VIREO
Vireo atricapillus   4½"
VOICE: A series of harsh, "angry" phrases in great variety. Alarm note: *chit-ah*.
HABITAT: Oak scrub, brushy hillsides. Limited to cen. Oklahoma, Edward's Plateau of cen. Texas, and Big Bend area of w. Texas.
NOTES: Very specific in habitat needs. Presently federally protected. Note red eye-ring.

VIREOS
With wingbars

sexes similar

YELLOW-THROATED
VIREO

WHITE-EYED VIREO

immature

BELL'S VIREO

BLUE-HEADED
VIREO

♂

♀

BLACK-CAPPED VIREO

WOOD-WARBLERS Family Parulidae

Small, brightly colored, active birds with thin bills. Majority have some yellow in plumage. Most leave the U.S. to winter on Caribbean Islands or in Cen. America. **FOOD:** Mainly insects, spiders, larvae.

NORTHERN PARULA *Parula americana* 4½"
VOICE: A buzzy trill that climbs the scale and then drops: *zeeeeeeeeeeee-up*.
HABITAT: Woodlands, often where *Usnea* lichen or Spanish moss dominates.
NOTES: Reestablishing breeding areas in some parts of the Northeast. Local.

"SUTTON'S" WARBLER
Hybrid between Yellow-throated and Parula. Known from 2 specimens and several sightings.

YELLOW-THROATED WARBLER
Dendroica dominica 5½"
VOICE: Clear, slurred notes that drop in pitch, with the last note rising: *tee-ew-tew-tew-tew wi*.
HABITAT: Open woods, live oaks, sycamores, pines.
NOTES: East Coast populations are mostly in pines while interior populations are closely associated with streamside sycamores.

BLACK-THROATED GREEN WARBLER
Dendroica virens to 5"
VOICE: Lazy *zee-zee-zoo-zee-zee* with the *zoo* at higher level.
HABITAT: Breeds in conifer and mixed forests. Visits variety of woodlands during migration.
NOTES: Loss of hemlocks in East may adversely affect some populations.

PROTHONOTARY WARBLER
Protonotaria citrea 5½"
VOICE: A loud, 1-pitched *zweet, zweet, zweet, zweet*.
HABITAT: Wooded swamps, river edges.
NOTES: A hole-nesting warbler that is expanding its range north and west.

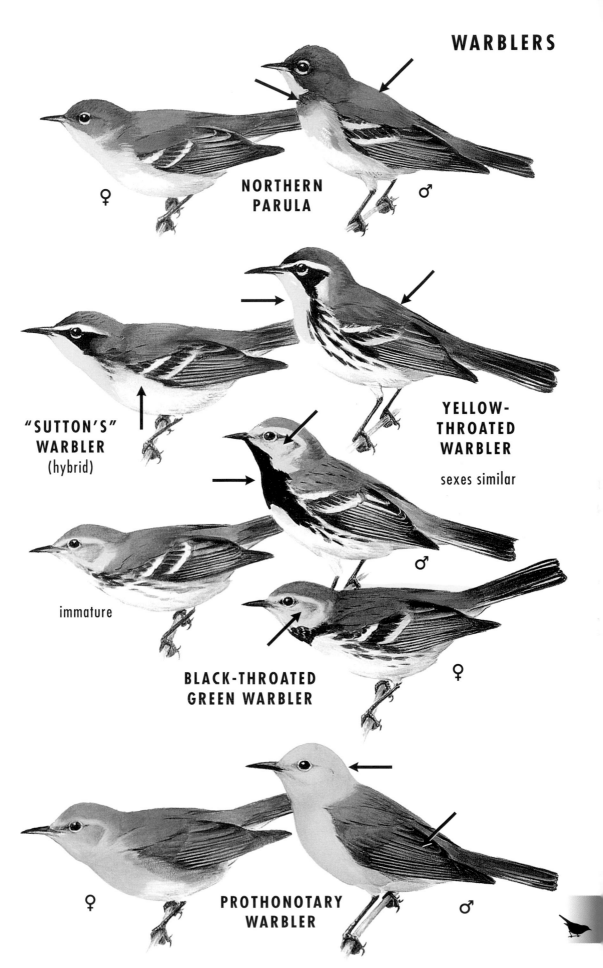

WARBLERS

NORTHERN PARULA

♀ ♂

"SUTTON'S" WARBLER (hybrid)

YELLOW-THROATED WARBLER

sexes similar

♂

immature

BLACK-THROATED GREEN WARBLER

♀

PROTHONOTARY WARBLER

♀ ♂

BLACK-AND-WHITE WARBLER
Mniotilta varia　　　　　　　　　　**to 5½"**

VOICE: A thin, pulsating *weesee, weesee, weesee, weesee*. Also series of jumbled notes.
HABITAT: Woodlands.
NOTES: Forages on tree trunks and limbs.

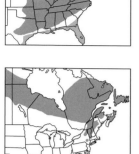

BLACKPOLL WARBLER
Dendroica striata　　　　　　　　　　**5"**

VOICE: A high-pitched, mechanical, pulsating *zi-zi-zi-zi-zi-zi-zi* stronger in middle.
HABITAT: Breeds in conifers. Woodlands in migration.
NOTES: In fall, may fly to S. America nonstop.

BLACK-THROATED GRAY WARBLER
Dendroica nigrescens　　　　　　　　　**5"**

Rare visitor from West. Winters with more regularity in s. Texas.

BLACK-THROATED BLUE WARBLER
Dendroica caerulescens　　　　　　　　**to 5½"**

VOICE: A wheezy *zur, zur, zur, zree, zree* that rises in pattern.
HABITAT: Understory of mixed woodlands.
NOTES: Feeds down low in forest.

CERULEAN WARBLER
Dendroica cerulea　　　　　　　　　　**to 4½"**

VOICE: Buzzy notes on 1 pitch with a sharp rise to the end notes: *zray, zray, zray, zreee*.
HABITAT: Mature woodlands, especially in river valleys.
NOTES: Population is declining overall, but there is some expansion in eastern states.

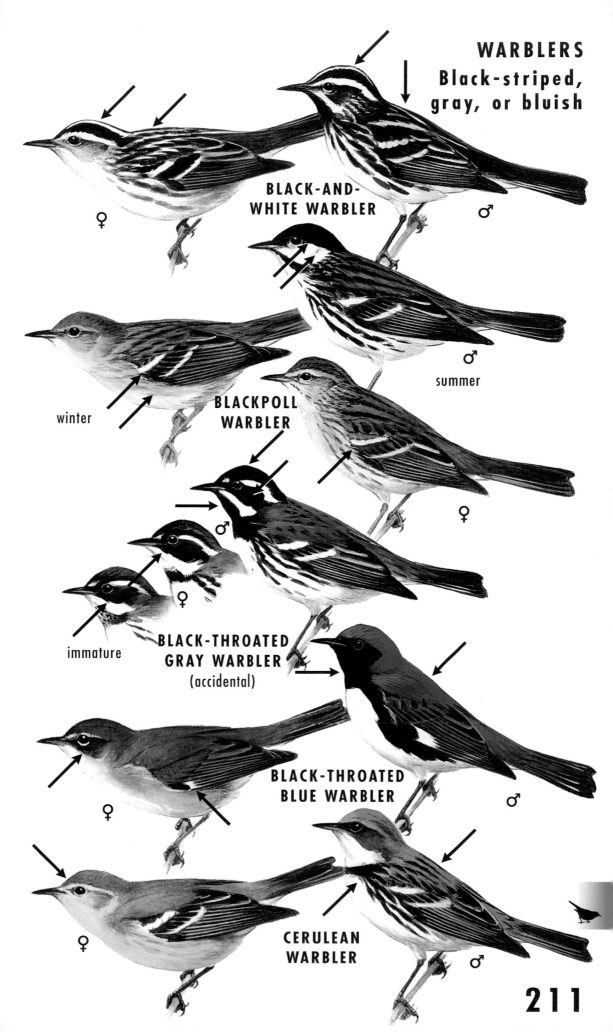

♀

BLACK-AND-
WHITE WARBLER

♂

♂

summer

winter

BLACKPOLL
WARBLER

♀

♂

♀

immature

BLACK-THROATED
GRAY WARBLER
(accidental)

♀

BLACK-THROATED
BLUE WARBLER

♂

♀

CERULEAN
WARBLER

♂

MAGNOLIA WARBLER
Dendroica magnolia   4⁷⁄₁₀"

VOICE: A loud *weeta, weeta, weetsee*, with the last note rising.

HABITAT: Nests in conifers. In migration any woodland or thicket.

NOTES: Sporadic breeding in conifer woods to the south of normal range.

YELLOW-RUMPED ("Myrtle") WARBLER
Dendroica coronata   6"

VOICE: A loose jumble of notes that rises in pitch, and drops at the end. Call note, loud *check*.

HABITAT: Conifer and mixed forests. Bayberry and scrub in winter areas.

NOTES: May winter well north. Western form ("Audubon's" Warbler) very rare in East.

KIRTLAND'S WARBLER
Dendroica kirtlandii   6"

VOICE: 3 to 4 loud staccato notes followed by higher-pitched ringing notes. Ends abruptly.

HABITAT: Confined to cen. Mich. jack pine woods that are 5–18 feet high. Winters in Bahamas.

NOTES: Breeding greatly hampered by Brown-headed Cowbird nest parasitism.

CANADA WARBLER
Wilsonia canadensis   to 5⁷⁄₁₀"

VOICE: A burst of irregularly arranged notes: *chip, chupety, swee-ditchety* (Gunn).

HABITAT: Forest undergrowth, shady thickets, swamplands.

NOTES: Winters in S. America.

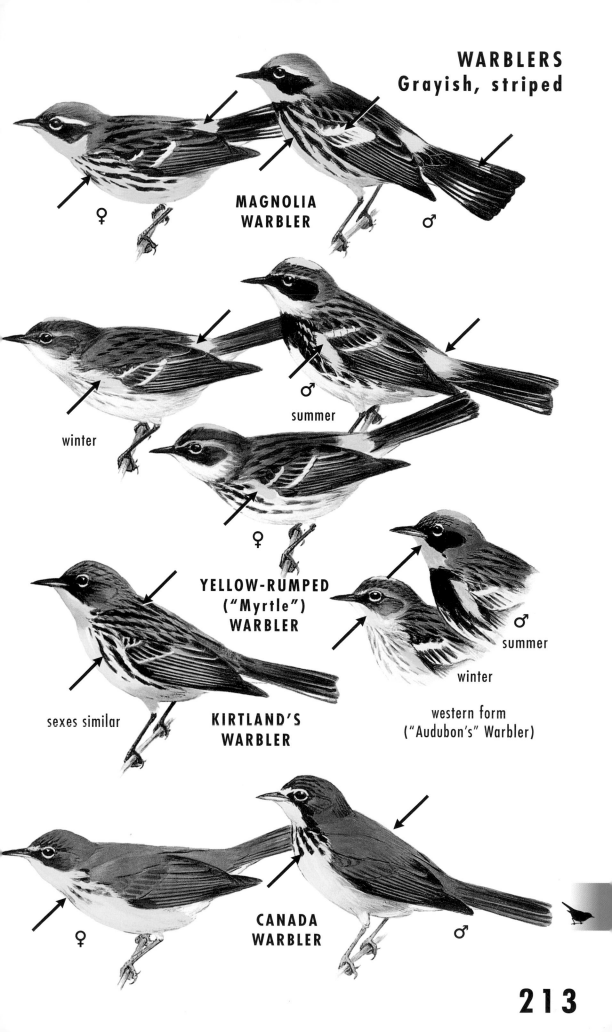

WARBLERS
Grayish, striped

MAGNOLIA
WARBLER

♀

♂

winter

♂

summer

♀

YELLOW-RUMPED
("Myrtle")
WARBLER

♂
summer

winter

sexes similar

KIRTLAND'S
WARBLER

western form
("Audubon's" Warbler)

CANADA
WARBLER

♀

♂

213

CAPE MAY WARBLER *Dendroica tigrina* 5"

VOICE: A high, thin *seet, seet, seet, seet.*

HABITAT: Nests in spruce forests; migrates through woodlands.

NOTES: Forages in the tops of trees and is often overlooked.

CHESTNUT-SIDED WARBLER

Dendroica pensylvanica **to 5½"**

VOICE: A distinct *pleeez pleeez ta meet cha.*

HABITAT: Pasture edges, thickets, lowlands.

NOTES: Note upcocked tail.

BAY-BREASTED WARBLER

Dendroica castanea **to 5½"**

VOICE: A high, sibilant *tees, teesi, teesi.*

HABITAT: Nests in conifers; visits woodlands in migration.

NOTES: Has spread into w. Nfld. Runs along limbs as it feeds.

BLACKBURNIAN WARBLER

Dendroica fusca **to 5"**

VOICE: A high *zip zip zip titi tseeee,* with the last note very high pitched. Also *teetsa, teetsa, teetsa zizizizizi.* End of songs inaudible to some ears.

HABITAT: Nests in coniferous or mixed forest; migrates through woodlands.

NOTES: Feeds high in trees. Male's chest is the color of an orange flame.

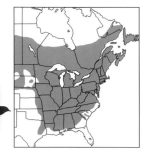

AMERICAN REDSTART

Setophaga ruticilla **to 5"**

VOICE: A loud, hurried *teetsa, teetsa, teetsa, tee-o.*

HABITAT: Woodlands, swamps, saplings.

NOTES: Active and acrobatic. Spreads tail while hopping about on limbs in zigzag pattern. Often flutters like a dropping butterfly when fly-catching.

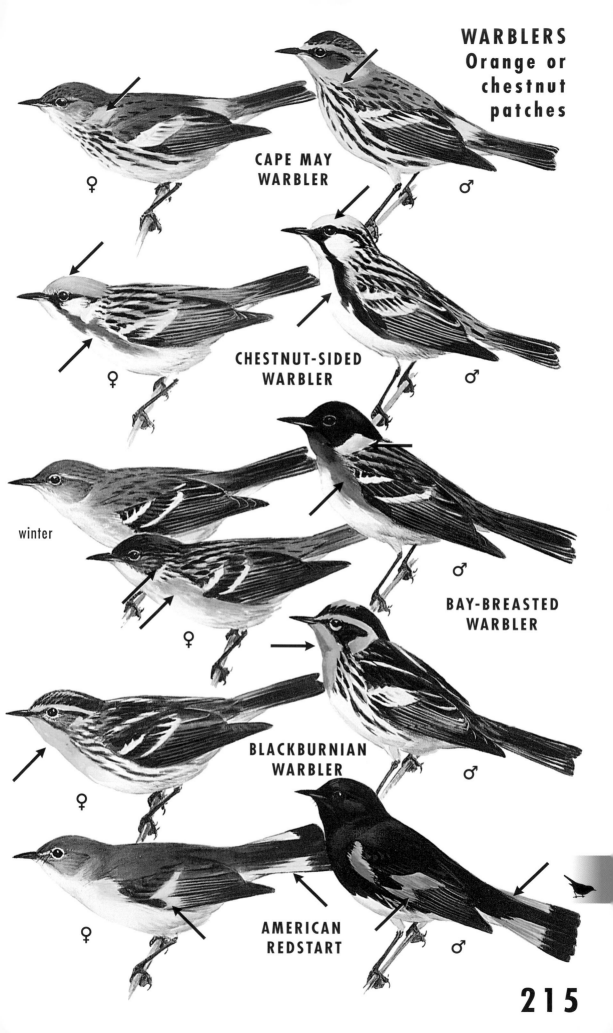

♀

♂

CAPE MAY
WARBLER

♀

♂

CHESTNUT-SIDED
WARBLER

winter

♂

BAY-BREASTED
WARBLER

♀

BLACKBURNIAN
WARBLER

♀

♂

♀

AMERICAN
REDSTART

♂

215

PINE WARBLER *Dendroica pinus* 5½"
VOICE: A musical, one-pitched trill.
HABITAT: Open pine woods and pine barrens.
NOTES: Increasing numbers in New England. Few may winter in the North. Earliest nesting warbler. Local in northern part of range.

PRAIRIE WARBLER *Dendroica discolor* 5"
VOICE: A thin *zee, zee, zee, zee, zee* up the scale.
HABITAT: Brushy pastures, power lines, scrub.
NOTES: A tail wagger that forages low. Numbers have increased in some areas, but it is often local because of specific habitat requirements.

PALM WARBLER *Dendroica palmarum* 5"
VOICE: A weak, repetitive *zhe, zhe, zhe, zhe.*
HABITAT: Nests at edge of muskeg. Visits woods, brushy, weedy, and other open areas during migration.
NOTES: Bobs tail. Most often is on or near the ground. Some may winter well north.

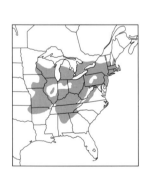

BLUE-WINGED WARBLER
Vermivora pinus 5"
VOICE: Buzzy *beee-buzz* (as if inhaled then exhaled).
HABITAT: Fields, brushy areas, wood edges.
NOTES: Expanding range, displacing Golden-winged.

"BREWSTER'S" WARBLER
Vermivora chrysoptera × pinus
Hybrid that occurs where Blue-winged and Golden-winged Warblers both occur. May sing like either parent. More common than "Lawrence's" hybrid (see p. 220).

YELLOW WARBLER *Dendroica petechia* 5"
VOICE: A rapid *weet weet weet tsee tsee.*
HABITAT: Swamp edges, wetlands, gardens, orchards, brushy fields.
NOTES: Subspecies "Golden" Warbler (not shown) is a resident in the Florida Keys.

immature

**PINE
WARBLER**

♂

♀

**PRAIRIE
WARBLER**

♂

"western" race

**PALM
WARBLER**

♂

winter

♂

yellowish race

hybrid
(Blue-wing ×
Golden-wing)

♂

**"BREWSTER'S"
WARBLER**

♂

**BLUE-WINGED
WARBLER**

♀

**YELLOW
WARBLER**

♂

217

SWAINSON'S WARBLER
Limnothlypis swainsonii   5"
VOICE: A loud, ringing, clear song with 2 slurred notes, 2 lower notes, and 1 higher note.
HABITAT: Wooded swamps in Miss. Lowlands and stream bottoms, swamps, and tangles.
NOTES: Casual as far north as Mass. Shy.

WORM-EATING WARBLER
Helmitheros vermivorus   to 5½"
VOICE: A thin, dry buzzy trill that resembles the song of a Chipping Sparrow. Rapid and insectlike.
HABITAT: Woodland slopes; ravines.
NOTES: Numbers have decreased in many areas. Very rarely spotted as far north as Me. and Maritimes.

TENNESSEE WARBLER
Vermivora peregrina   4⁷⁄₁₀"
VOICE: A staccato *ticka ticka ticka ticka swit swit, chew-chew-chew.*
HABITAT: Mixed forests, bog edges; wide variety of woodlands during migration.
NOTES: One of several species of northern-breeding warblers whose population is somewhat cyclical in response to spruce budworm outbreaks.

ORANGE-CROWNED WARBLER
Vermivora celata   to 5"
VOICE: A colorless trill that grows weaker at the end. Often changes in pitch, rising and then dropping.
HABITAT: Brushy clearings, aspens, undergrowth, thickets, weedy tangles.
NOTES: Seen as a migrant. More sightings occur in the fall. Drab and often overlooked. Casual in winter as far north as Cape Cod.

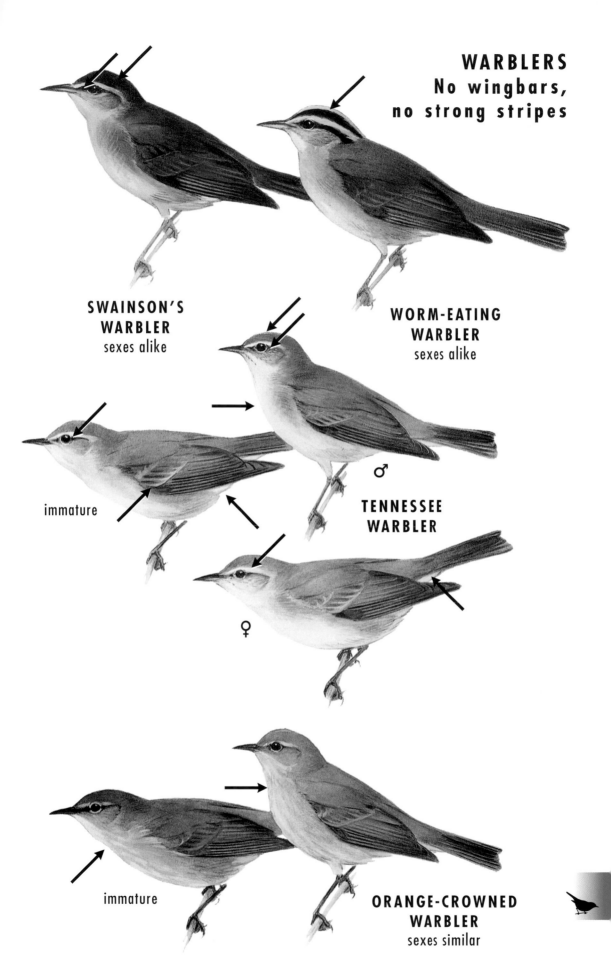

WARBLERS
No wingbars, no strong stripes

SWAINSON'S WARBLER
sexes alike

WORM-EATING WARBLER
sexes alike

immature

♂

TENNESSEE WARBLER

♀

immature

ORANGE-CROWNED WARBLER
sexes similar

WILSON'S WARBLER *Wilsonia pusilla* 4⁷⁄₁₀"

VOICE: A series of notes that drops in pitch at the end: *chi chi chi chi chet chet.*

HABITAT: Thickets along streams, brushy tangles, willows, and alders.

NOTES: A northern breeder that is always very active. Flits wings and flicks tail.

HOODED WARBLER *Wilsonia citrina* 5½"

VOICE: A loud, whistled *wheeta wheeta tee-o.*

HABITAT: Woodland undergrowth (prefers laurels). Wooded swamps.

NOTES: Stays near ground. Casual as far north as Nova Scotia.

BACHMAN'S WARBLER

Vermivora bachmanii 4²⁄₅"

Probably extinct: has not been reported since 1962. Lived in southern cypress swamplands with cane thickets. Song was a mechanical buzz on one pitch. Also a twittering buzzy song given in flight (Proctor/Petersen).

GOLDEN-WINGED WARBLER

Vermivora chrysoptera 5½"

VOICE: A buzzy *bee-bz-bz-bz;* the last 3 notes are usually on a lower scale.

HABITAT: Brushy fields, undergrowth, field edges, power lines.

NOTES: Being displaced by the Blue-winged Warbler in many areas.

"LAWRENCE'S" WARBLER

Vermivora chrysoptera × pinus

The recessive and therefore rarer hybrid between Blue-winged and Golden-winged Warblers. The appearance of this species usually heralds the end of Golden-wings in that area. Occurs in the same habitat and may sing like either parent.

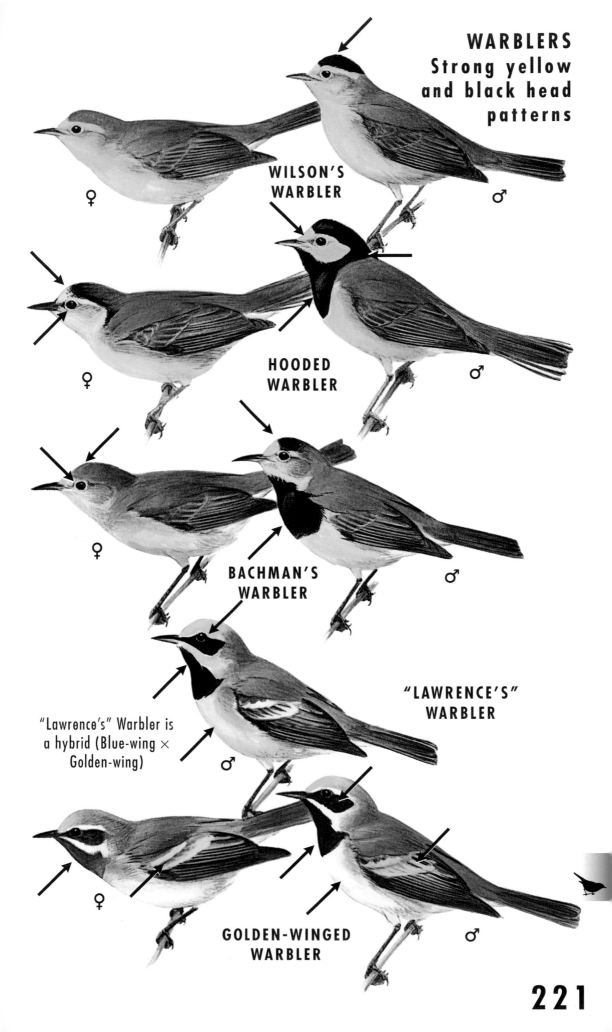

WARBLERS
Strong yellow
and black head
patterns

WILSON'S
WARBLER

♀

♂

HOODED
WARBLER

♀

♂

BACHMAN'S
WARBLER

♀

♂

"LAWRENCE'S"
WARBLER

"Lawrence's" Warbler is
a hybrid (Blue-wing ×
Golden-wing)

♂

♀

GOLDEN-WINGED
WARBLER

♂

221

NASHVILLE WARBLER

Vermivora ruficapilla   4 ⁷⁄₁₀"

VOICE: A 2-parted *seebit, seebit, seebit, seebit, titititi-tititi.*

HABITAT: Forest edges; bogs; open mixed woods with undergrowth.

NOTES: Nests sporadically in lower edge of its range. Often found high in trees during migration.

CONNECTICUT WARBLER

Oporornis agilis   **to 6"**

VOICE: A repeated *chip-chup-ee, chip-chup-ee* or *sugar-tweet, sugar-tweet* (W. Gunn).

HABITAT: Poplar bluffs, muskeg, woods near water. Visits Impatiens and Clethra thickets in migration.

NOTES: Spring migrant in Mississippi River Valley. Fall migrant on Atlantic Coast. Walks rather than hops. Can be shy and difficult to see.

MOURNING WARBLER

Oporornis philadelphia   5 ⁷⁄₁₀"

VOICE: A ringing *chirry, chirry, chorry, chorry,* with the *chorry* notes lower.

HABITAT: Clearings, thickets, slashings.

NOTES: Moves primarily northwest of the Appalachians and through the Ohio River Valley.

KENTUCKY WARBLER

Oporornis formosus   5 ½"

VOICE: A rapid, rolling chant of *tory, tory, tory.*

HABITAT: Moist woodland undergrowth, laurel thickets.

NOTES: Casual north as far as Nova Scotia. Ten Kentuckys are heard for every one seen.

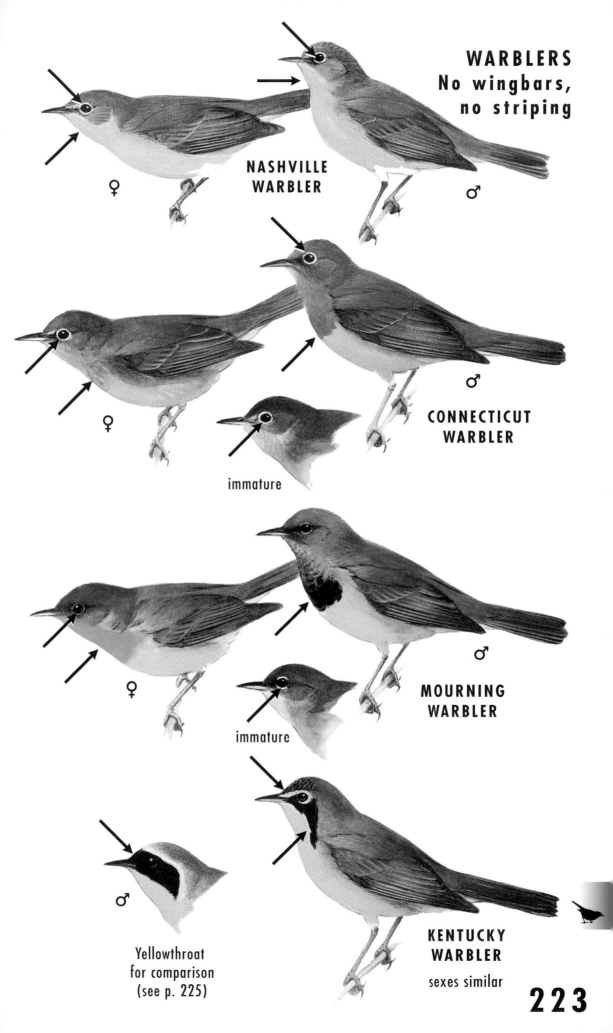

WARBLERS
No wingbars, no striping

NASHVILLE
WARBLER

♀

♂

CONNECTICUT
WARBLER

♀

♂

immature

MOURNING
WARBLER

♀

♂

immature

♂

Yellowthroat
for comparison
(see p. 225)

KENTUCKY
WARBLER

sexes similar

223

COMMON YELLOWTHROAT
Geothlypis trichas   **to 5"**
VOICE: A rapid chant: *witchy, witchy, witchy* or *witchity, witchity witch.*
HABITAT: Swamps, marshes, wet thickets. A skulker.
NOTES: In winter, seen rarely as far north as Nova Scotia.

YELLOW-BREASTED CHAT *Icteria virens* **7"**
VOICE: Clear, repeated harsh whistles and cackles. Long pauses between phrases.
HABITAT: Brushy tangles, briars and thickets, grape vines. Open fields with scattered brush clumps.
NOTES: Disappearing in areas at northern edge of habitat. A few winter in U.S.

NORTHERN WATERTHRUSH
Seiurus noveboracensis   **6"**
VOICE: Note is a sharp, metallic *chink*. Song is a clear, whistled *twit twit twit sweet sweet sweet chew chew chew. Chews* drop in pitch.
HABITAT: Swampy or wet woods, stream edges, lakeshores.
NOTES: Walks and bobs tail. In migration its *chink* note gives it away in thickets.

LOUISIANA WATERTHRUSH
Seiurus motacilla   **6"**
VOICE: Song a jumble of ringing musical notes starting with 3 clear, slurred whistles.
HABITAT: Brooks, ravines, stream edges.
NOTES: Walks and bobs tail. Early spring arrival (and early departure) in the Northeast.

OVENBIRD *Seiurus aurocapillus* **6"**
VOICE: An emphatic and rising *teach-er, teach-er, teach-er!*
HABITAT: Near ground or walking on ground in woodlands and thickets.
NOTES: There have been winter records from as far north as Mass.

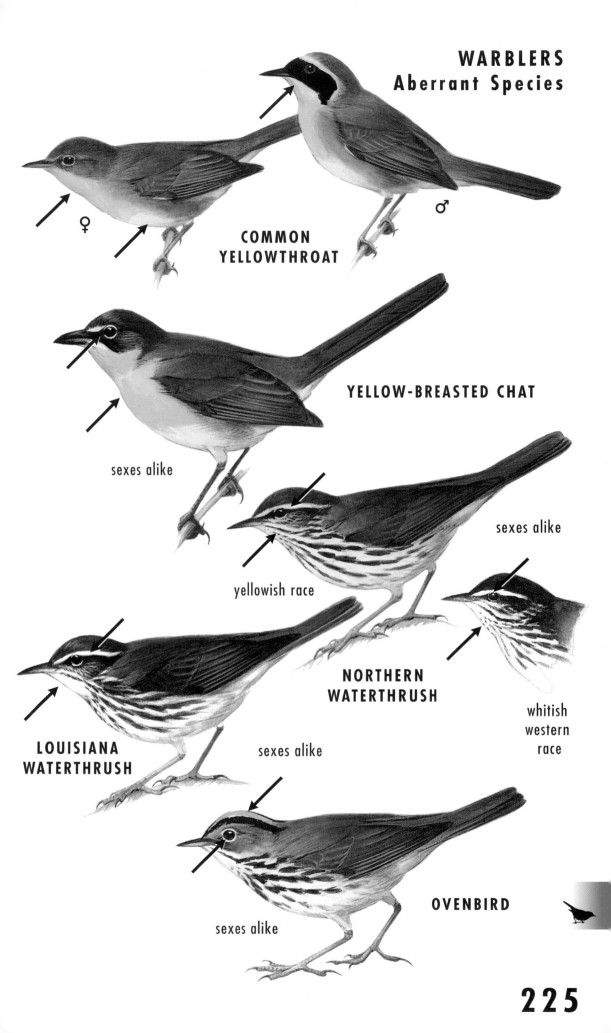

WARBLERS
Aberrant Species

COMMON
YELLOWTHROAT

♀

♂

YELLOW-BREASTED CHAT

sexes alike

sexes alike

yellowish race

NORTHERN
WATERTHRUSH

LOUISIANA
WATERTHRUSH

whitish
western
race

sexes alike

OVENBIRD

sexes alike

CONFUSING FALL WARBLERS

Most of these birds have streaks or wingbars.

RUBY-CROWNED KINGLET *Regulus calendula* **p. 194**
(Not a warbler.) Broken eye-ring. Flicks wings.

CHESTNUT-SIDED WARBLER *Dendroica pensylvanica* **p. 214**
Immature: Yellow-green above, whitish below; eye-ring. Cocked tail.

BAY-BREASTED WARBLER *Dendroica castanea* **p. 214**
Note dark legs, buff undertail coverts. Fall adult may have bay on flanks.

BLACKPOLL WARBLER *Dendroica striata* **p. 210**
To distinguish from look-alikes, look for a greenish tint below with streaks, streaked back. White undertail coverts and pale yellowish feet.

PINE WARBLER *Dendroica pinus* **p. 216**
Note white undertail coverts, black legs. Unstreaked back. Dark cheek.

NORTHERN PARULA *Parula americana* **p. 208**
Immature: Combination of bluish and yellow; wingbars.

MAGNOLIA WARBLER *Dendroica magnolia* **p. 212**
Immature: White band at midtail.

PRAIRIE WARBLER *Dendroica discolor* **p. 216**
Immature: Jaw stripe, side stripes. Bobs tail.

YELLOW WARBLER *Dendroica petechia* **p. 216**
Yellow tail spots. Short tail. Beady dark eye. Yellow edging to wing.

BLACKBURNIAN WARBLER *Dendroica fusca* **p. 214**
Immature: Yellow throat, dark cheek; pale back stripes.

BLACK-THROATED GREEN WARBLER *Dendroica virens* **p. 208**
Immature: Yellow frames a pale olive cheek. Green back.

PALM WARBLER *Dendroica palmarum* **p. 216**
Brownish back, yellowish undertail coverts. Bobs tail.

YELLOW-RUMPED ("Myrtle") WARBLER *Dendroica coronata* **p. 212**
Immature: Bright yellow rump. Yellow spot at side of breast.

CAPE MAY WARBLER *Dendroica tigrina* **p. 214**
Streaked breast, dull yellowish rump. Note the pale neck spot.

CONFUSING FALL WARBLERS

♀
RUBY-CROWNED KINGLET
not a warbler

immature
CHESTNUT-SIDED

adult
PINE

BAY-BREASTED

immature

immature

immature
PINE

immature
BLACKPOLL

immature
NORTHERN PARULA

immature
MAGNOLIA

immature
PRAIRIE

immature

immature

immature
BLACKBURNIAN

immature
BLACK-THROATED GREEN

YELLOW

immature
CAPE MAY

PALM

immature
YELLOW-RUMPED ("Myrtle")

227

CONFUSING FALL WARBLERS

Most of these birds lack streaks or wingbars.

ORANGE-CROWNED WARBLER *Vermivora celata*   p. 218
Dingy breast, yellow undertail coverts. Fall immature is greenish drab throughout, barely paler below; it is often quite gray.

TENNESSEE WARBLER *Vermivora peregrina*   p. 218
Trace of a wingbar; white undertail coverts. Greenish upperparts.

PHILADELPHIA VIREO *Vireo philadelphicus*   p. 204
(Not a warbler.) "Vireo" bill and actions (see p. 204).

HOODED WARBLER *Wilsonia citrina*   p. 220
Immature: Yellow eyebrow stripe, bold white tail spots.

WILSON'S WARBLER *Wilsonia pusilla*   p. 220
Immature: Dusky forehead, no white in its longish tail.

BLACK-THROATED BLUE WARBLER
Dendroica caerulescens   p. 210
Dark-cheeked look, white wing spot. Some young birds and females lack this white or light "pocket handkerchief" and may suggest Philadelphia Vireo or Tennessee Warbler (above), but note the dark cheek.

CONNECTICUT WARBLER *Oporornis agilis*   p. 222
Immature: Suggestion of a hood; complete circular eye-ring. Walks.

MOURNING WARBLER *Oporornis philadelphia*   p. 222
Immature and fall female: Suggestion of a hood; yellowish throat; broken eye-ring. Brighter yellow below than Connecticut.

NASHVILLE WARBLER *Vermivora ruficapilla*   p. 222
Yellow throat, white eye-ring, grayish head.

COMMON YELLOWTHROAT *Geothlypis trichas*   p. 224
Female: Yellow throat, brownish sides, white belly.

PROTHONOTARY WARBLER *Protonotaria citrea*   p. 208
Female: Dull golden head, gray wings.

CANADA WARBLER *Wilsonia canadensis*   p. 212
Immature: Yellow and white "spectacles," trace of necklace.

CONFUSING FALL WARBLERS

immature

ORANGE-CROWNED

TENNESSEE

PHILADELPHIA
VIREO
not a warbler

WILSON'S

HOODED

♀

immature

BLACK-THROATED
BLUE

♀

immature

CONNECTICUT

MOURNING

immature

NASHVILLE

♀

COMMON
YELLOWTHROAT

♀

PROTHONOTARY

immature

CANADA

A varied group possessing conical, sharp-pointed bills and rather flat profiles. Some are black and iridescent; others are highly colored. **FOOD:** Insects, small fruits, seeds, waste grain, small aquatic life.

RED-WINGED BLACKBIRD
Agelaius phoeniceus **to 9½"**
VOICE: Call note is a loud *check* and high, slurred whistle: *terrr-eeee*. Song is a liquid *o-ka-lee*.
HABITAT: Marshes, swamps, hay fields, cultivated lands.
NOTES: Very gregarious. The voice of the spring marsh. Male's brilliant epaulets flare in display.

YELLOW-HEADED BLACKBIRD
Xanthocephalus xanthocephalus **to 11"**
VOICE: An electronic buzzing with interspersed croaks, whistles, and rasping notes.
HABITAT: Freshwater marshes. Forages in cultivated fields and open country.
NOTES: A prairie species that occasionally appears in grackle and blackbird flocks east to the East Coast, typically in the fall and winter.

BROWN-HEADED COWBIRD
Molothrus ater **to 7"**
VOICE: Flight call is a *weeee-titi*. Song is a gurgling, bubbling, upward *glug glug gleee*.
HABITAT: Farms, fields, roadsides, parks, wood edges.
NOTES: Brown-headed Cowbirds have had a distinct impact on some species, as they are nest parasites, leaving eggs in other birds' nests. The more edge created, the more woodland birds are affected. Note high cock to tail when feeding in mixed blackbird flocks.

ICTERIDS
(BLACKBIRDS, etc.)

epaulets
concealed

immature ♂

♀

epaulets in
display

♂

♂

RED-WINGED BLACKBIRD

♀

YELLOW-HEADED
BLACKBIRD

♂

♀

BROWN-HEADED
COWBIRD

juvenile

immature ♂ molting

RUSTY BLACKBIRD *Euphagus carolinus* 9"
VOICE: Call note is a sharp *chack*. Song is a split creak like a rusty hinge: *kush-a-lee*.
HABITAT: Riverine woods, wooded swamps.
NOTES: Male attains rusty marks only in the fall. Forages for food on ground, flipping wet leaves.

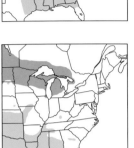

BREWER'S BLACKBIRD
Euphagus cyanocephalus to 9"
VOICE: A harsh, wheezy, creaking *queee-ee*.
HABITAT: Fields, prairies, farms, parks.
NOTES: A scarce visitor to the East Coast.

COMMON GRACKLE
Quiscalus quiscula to 13½"
VOICE: Split rasping notes with calls of *chack*.
HABITAT: Croplands, woodlands, towns, evergreen groves.
NOTES: Gregarious; forages in huge flocks. The two color races, "purple" and "bronze," were considered separate species at one time.

BOAT-TAILED GRACKLE
Quiscalus major to 16½"
VOICE: A jumble of harsh clicks, buzzes, and whistles. Large roosting flocks can be deafening.
HABITAT: Near salt marshes along coast and fresh marshes. Croplands.
NOTES: Extending range north. Overlaps with Great-tailed Grackle in coastal Texas and Louisiana.

GREAT-TAILED GRACKLE
Quiscalus mexicanus to 18"
VOICE: Buzzes, clicks, and cackles like the Boat-tailed. Also sounds a harsh *check check*.
HABITAT: Coastal areas, marshes, parks, towns.
NOTES: Male has flatter head and larger tail than male Boat-tailed Grackle. Extending range north.

ICTERIDS
(BLACKBIRDS, etc.)

♂ summer

♀

♂ winter

RUSTY
BLACKBIRD

BREWER'S
BLACKBIRD

♀

♂ winter

♂ summer

♂

♂

purple race

♀

bronze
race

COMMON
GRACKLE

Atlantic
Coast

♀

♂

♂

♀

GREAT-
TAILED
GRACKLE

♂

Gulf
states

BOAT-TAILED
GRACKLE

233

BOBOLINK *Dolichonyx oryzivorus*   to 8"
VOICE: Clear, bubbling, reedy notes are often given in flight while hovering. Flight note a clear *pink*.
HABITAT: Hay fields, meadows.
NOTES: Population is declining in many areas because of diminishing grasslands and harvesting schedules.

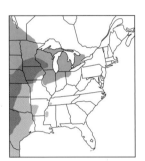

EASTERN MEADOWLARK
Sturnella magna   9"
VOICE: 2 clear, slurred whistles that drop at the end: *tee-yah tee-yair*. Buzzy *dzrrt*.
HABITAT: Fields, meadows, prairies, marsh edges.
NOTES: Numbers are decreasing in many areas because of loss of habitat.

WESTERN MEADOWLARK
Sturnella neglecta   9"
VOICE: 7–10 flutelike and gurgling notes introduced by 1 clear whistle.
HABITAT: Fields, meadows, prairies.
NOTES: Although it overlaps with the Eastern Meadowlark, it rarely interbreeds. The eastern range in winter is not well known.

STARLINGS Family Sturnidae

A family of more than 100 species that vary in color from iridescent purples to brilliant oranges and yellows. Highly gregarious. Our one species was introduced from Europe. **FOOD:** Everything!

EUROPEAN STARLING
Sturnus vulgaris   to 8½"
VOICE: A jumble of whistles, squeaks, and slurred notes. Occasionally attempts to mimic other birds.
HABITAT: Has adapted to live in areas developed by humans. Not a deep woodland species.
NOTES: A problem species that has impacted native species such as the Eastern Bluebird, from which it takes its nesting hole. Large roosting congregations cause problems.

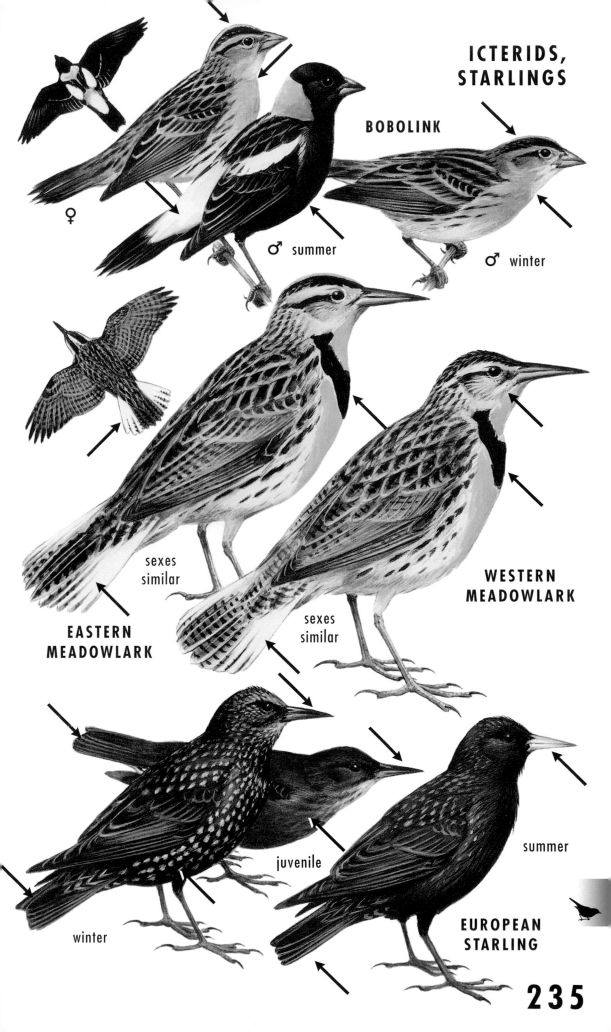

ICTERIDS,
STARLINGS

BOBOLINK

♀

♂ summer

♂ winter

EASTERN
MEADOWLARK

sexes
similar

WESTERN
MEADOWLARK

sexes
similar

juvenile

summer

winter

EUROPEAN
STARLING

ORIOLES Family Icteridae

Brightly colored members of the Blackbird family. **FOOD:** Insects, spiders, fruit, nectar.

ORCHARD ORIOLE *Icterus spurius*   **to 7"**
VOICE: A fast outburst of clear whistles that bubble to an ending of *wheeer*. Call a soft *chuck*.
HABITAT: Wood edges, orchards, shade trees.
NOTES: Expanding in northern part of its range. First-year males in yellow plumage have black throat.

BALTIMORE ORIOLE *Icterus galbula*   **to 8"**
VOICE: Three clear whistles, then a bubbling *twee-dle-eet-doot*.
HABITAT: Open woods, parks, shade trees, river edges. Common in towns.
NOTES: Its pendulous white nest is a familiar sight. A few winter at northern feeders.

BULLOCK'S ORIOLE
Icterus bullockii   **to 8½"**
VOICE: A series of double notes followed by 1 or 2 piping notes. Also a sharp *skip*; a rough chatter.
HABITAT: Riverine forests and woodlands.
NOTES: Stragglers occur at feeders in the East in winter.

SPOT-BREASTED ORIOLE
Icterus pectoralis   **8"**
VOICE: A loud, long ensemble of whistles—classicly oriole-like.
HABITAT: Shade trees, parks, gardens. Introduced.
NOTES: Found only on east coast of Florida. The only spot-breasted oriole in the area.

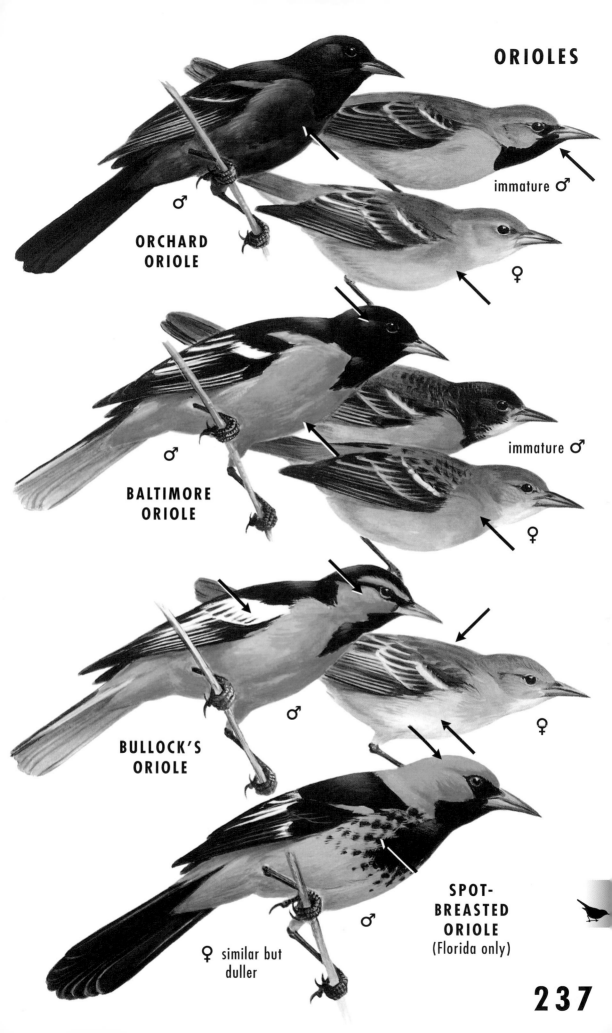

ORIOLES

ORCHARD ORIOLE

♂

immature ♂

♀

BALTIMORE ORIOLE

♂

immature ♂

♀

BULLOCK'S ORIOLE

♂

♀

SPOT-BREASTED ORIOLE
(Florida only)

♂

♀ similar but duller

237

Male tanagers are brightly colored; females are olive green above, yellowish below. Thick bills. Mainly tropical birds. A few species reach the U.S. FOOD: Insects, fruit.

SUMMER TANAGER *Piranga rubra* to 7⁷⁄₁₀"

VOICE: Call note is a staccato *pik-i-tuk-i-tuk*. Song is robinlike and clear; questioning.
HABITAT: Woodlands (especially oaks), river edges.
NOTES: Casual in spring as far north as Nova Scotia. Rarely seen at winter feeders.

SCARLET TANAGER *Piranga olivacea* 7"

VOICE: A raspy, see-sawing *tweer, tawoo, tweer*. Call note *chip-burrr*.
HABITAT: Woodlands (especially oaks), shade trees, hillsides.
NOTES: Males in changing plumage have blotches of red or orange.

WESTERN TANAGER *Piranga ludoviciana* 7"

A western species that very rarely wanders east. Often visits winter feeders. Use care in identification, as young Scarlet Tanagers can have a thin yellowish wingbar. Check for dusky saddleback appearance.

BLUE-GRAY TANAGER
Thraupis episcopus 6"

Shown here based on a population that at one time was established in and around Hollywood, Florida. There have been no recent sightings.

TANAGERS

SUMMER TANAGER

♀

♂ all seasons

immature changing to adult

♀

♂ changing

♂ winter

SCARLET TANAGER

♂ summer

♂ orange variant

WESTERN TANAGER

♀

♂ summer

♂ winter

BLUE-GRAY TANAGER
sexes similar

OLD WORLD SPARROWS Family Passeridae

Two representatives of an Old World family. Brought to U.S. from England. Not related to native sparrows. **FOOD:** Insects and seeds.

HOUSE SPARROW *Passer domesticus*   6"
VOICE: A jumble of unstructured notes. Squeaks, twitters, and whistles.
HABITAT: Cities, towns, and farms.
NOTES: Introduced. Has moved into locales in which humans have modified the environment.

EURASIAN TREE SPARROW
Passer montanus   5³/₅"
VOICE: A repeated *chit-tchup* and a jumble of higher notes. Flight note is a distinct *tek, tek*.
HABITAT: Farmlands and waste places.
NOTES: Introduced in St. Louis in 1870. Only recently has shown some spreading north into other areas.

GROSBEAKS, FINCHES, SPARROWS, BUNTINGS
Families Fringillidae, Emberizidae, Cardinalidae

Seed-cracking bills range from massive in grosbeaks to conical in sparrows to cross-tipped in crossbills. **FOOD:** Seeds, fruits, insects.

DICKCISSEL *Spiza americana*   to 7"
VOICE: A staccato *dick-ciss-ciss-ciss*. A rough buzzing note is heard overhead in migration.
HABITAT: Fields, meadows, prairies.
NOTES: Some wander east in fall and winter and may visit feeding stations. Often occurs with House Sparrows. Populations often move from place to place. Sporadic nesting east to dash line.

LARK BUNTING
Calamospiza melanocorys   7"
VOICE: Cardinal-like slurs, chips, and whistles.
HABITAT: Plains and prairies.
NOTES: Sporadic at edges of range. Casual or accidental in East.

FINCHES AND FINCHLIKE BIRDS

HOUSE SPARROW

♂

♀

EURASIAN TREE SPARROW

sexes similar

DICKCISSEL

♂

♀

fall

Bobolink for comparison (see p. 234)

♂ summer

winter ♂ similar to ♀

LARK BUNTING

♀

LAPLAND LONGSPUR
Calcarius lapponicus   **6½"**

VOICE: A musical tinkling that cascades downward. Call note is a hard, dry rattle; also a soft *tew*.

HABITAT: Breeds in tundra. A winter visitor to fields, prairies, and shoreline.

NOTES: Sparrowlike in winter plumage; hugs ground while feeding.

CHESTNUT-COLLARED LONGSPUR
Calcarius ornatus   **6½"**

VOICE: Song is short and musical but feeble. Note is a finchlike *kittle-kittle*.

HABITAT: Plains and prairies.

NOTES: Accidental to the East Coast. Note tail patterns in longspurs for identification.

McCOWN'S LONGSPUR
Calcarius mccownii   **6"**

VOICE: Clear sweet warbles are delivered in display flight. Call note is a dry rattle.

HABITAT: Plains and prairies.

NOTES: Range has withdrawn westward. Accidental birds have reached east to Mass.

SMITH'S LONGSPUR   *Calcarius pictus*   **6"**

VOICE: Sweet and warbling, ending in *we-chew*. Call notes are rattles and clicks (suggesting the winding of a cheap watch!).

HABITAT: Breeds in Arctic tundra and boreal edge. Winter: fields, prairies, airports.

NOTES: Accidental to the East Coast. It is somewhat similar in appearance to the Vesper Sparrow, but its shoulder patch is white rather than chestnut and it is buffier and has less streaking below.

LONGSPURS

♀ winter

LAPLAND
LONGSPUR

♂ winter

♂ summer

CHESTNUT-
COLLARED
LONGSPUR

♀

♂ winter

♂ summer

McCOWN'S
LONGSPUR

♀

♂ winter

♂ summer

♀

♂ winter

See Horned Lark
and pipits, p. 179

SMITH'S
LONGSPUR

♂ summer

243

DARK-EYED ("Slate-colored") JUNCO
Junco hyemalis   to 6½"
VOICE: A loose, musical trill like that of a Pine Warbler or Chipping Sparrow, but smoother.
HABITAT: Breeds in conifer and mixed woods. Winters in open woods, undergrowth, brush, feeders.
NOTES: This species is a common visitor to the winter feeder.

DARK-EYED ("Oregon") JUNCO
Junco hyemalis   to 6½"
VOICE: Similar to that of the "Slate-colored."
HABITAT: Western form that turns up rarely at feeding stations in the East.
NOTES: The "White-winged" form (not shown) breeds in the Black Hills of South Dakota. Several other western forms winter along western edge of map.

SNOW BUNTING
Plectrophenax nivalis   to 7⅕"
VOICE: A musical *ti-ti-chu-ree*. Call note is a sharp whistled *teer* or *tew*, also a rolling *brrreeet*.
HABITAT: Breeds in Arctic tundra. Wintering birds visit prairies, fields, farmland, and the shoreline.
NOTES: No other songbird shows so much white, but beware of albino or leucistic individuals of other species. Often overlooked, as it blends in with snow and grasses.

"SNOWBIRDS"

DARK-EYED
("Slate-colored")
JUNCO

♀

juvenile

♂

DARK-EYED
("Oregon")
JUNCO

♂

♀

♀ winter

♂ winter

♂
summer

SNOW
BUNTING

245

NORTHERN CARDINAL
Cardinalis cardinalis   **to 9"**
VOICE: A clear, whistled *what-cheer, what-cheer* followed by *whoit, whoit, whoit*. Call note is a sharp, metallic *chip*.
HABITAT: Woodland edges, thickets, gardens, parks, lowlands.
NOTES: One of the best known birds because of its color. Expanding its range north. A common visitor to feeders.

RED CROSSBILL *Loxia curvirostra*   **to 6½"**
VOICE: A finchlike warble: *jip, jip, jip, jeeaa, jeeaa*. Call note is a *jip-jip* or *kip-kip*.
HABITAT: Coniferous forests of the North and higher elevations. Nomadic, invading to the south at irregular intervals as far as dash line.
NOTES: Bill is unique to crossbills; it is used to open evergreen cones. Flocks visiting from North are often tame. Dangles from cones as it feeds.

WHITE-WINGED CROSSBILL
Loxia leucoptera   **to 6⁷⁄₁₀"**
VOICE: A succession of loud trills at different pitches. Call note is a liquid *peet* or dry *chif-chif*.
HABITAT: Spruce and fir forests. Invades to the south at irregular intervals as far as dash line.
NOTES: Particularly fond of hemlocks. Very rarely visits feeders.

RED FINCHES

NORTHERN
CARDINAL

♂

♀

juvenile

♂

♀

RED
CROSSBILL

immature

♂

♀

WHITE-WINGED
CROSSBILL

COMMON REDPOLL *Carduelis flammea* **5"**
VOICE: A trill followed by a rattling *chet-chet-chet*.
Flight call: a rattling *chet-chet-chet-chet*.
HABITAT: Breeds in tundra scrub. In summer wanders to birches, weeds, brush, shrubs.
NOTES: An erratic winter invader from the North to dash line. Often visits feeders with goldfinch flocks.

HOARY REDPOLL *Carduelis hornemanni* **5"**
VOICE: Calls and song like Common Redpoll's.
HABITAT: Same areas as that of Common Redpoll.
NOTES: A frosty white bird. Bill is smaller than Common Redpoll's. Much rarer wanderer to northern states south to dash line.

HOUSE FINCH
Carpodacus mexicanus **to 5⁷/₁₀"**
VOICE: A bright, loose, disjointed series of sweet notes that often ends with *wheer*.
HABITAT: Cities, suburbs, farms, parks.
NOTES: A western species introduced to the East in 1940s. Colonization of U.S. is nearly complete. Flocks often take over bird feeders.

PURPLE FINCH
Carpodacus purpureus **to 6"**
VOICE: A fast, lively warbler with jumbled sweet notes. Flight note: *pik*.
HABITAT: Woodlands; travels to suburbs in winter. Numbers in fall and winter vary.
NOTES: Appears to have been ousted from feeders by invading House Finches.

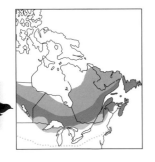

PINE GROSBEAK *Pinicola enucleator* **to 10"**
VOICE: A whistled *tee-tew-tew* like that of a yellowlegs but low-pitched.
HABITAT: Breeds in conifer forests. Wanders south into mixed woods and edges.
NOTES: A sporadic winter visitor south to dash line. Often very tame.

RED FINCHES, etc.

♀

♂

orange
variant

♀

♂ HOUSE
FINCH

COMMON
REDPOLL

♂

HOARY
REDPOLL

♀

♂ PURPLE
FINCH

immature ♂

♀

♂

PINE GROSBEAK

249

EVENING GROSBEAK
Coccothraustes vespertinus   **8"**

VOICE: A finchlike warble. Also ringing finchlike calls: *cleer-cleer* or *cleer-ip*.

HABITAT: Breeds in conifer and mixed forests. In winter to woodlands, field edges, feeders.

NOTES: An exotic-looking bird. Often works feeders through a winter. Irregular; major movements some years south to dash line.

AMERICAN GOLDFINCH   *Carduelis tristis*   **5"**

VOICE: Sustained, clear, and canary-like. Flight call: *per-chick-o-ree*.

HABITAT: Areas of thistle, lawns, weed patches, fields, and open woods.

NOTES: Dramatic plumage change from summer to winter. Eats thistle seeds at feeder.

PINE SISKIN   *Carduelis pinus*   **to 5"**

VOICE: A goldfinchlike song. Call notes are a loud *chleee-ip* or a high upward-slurred *shreeeee*.

HABITAT: Conifers, mixed woods, alder thickets, and weedy areas. During winter, often visits feeders with goldfinch groups.

NOTES: Sporadic in its movements from more northern areas. Following invasion years, may breed south to dash line.

EUROPEAN GOLDFINCH
Carduelis carduelis   **5½"**

This Eurasian introduction established itself for a short while on Long Island. The colony is now extirpated. Escaped birds occasionally appear at feeders.

YELLOW
FINCHES,
etc.

♂

♀

EVENING
GROSBEAK

♂ winter

♂ summer

♀

AMERICAN
GOLDFINCH

♂

♀

PINE SISKIN

sexes similar

EUROPEAN GOLDFINCH

251

BLUE GROSBEAK *Guiraca caerulea* **to 7½"**
VOICE: A rapid warble of short phrases that rise and fall. Call note is a metallic *chink*.
HABITAT: Brushy pastures, roadsides, stream edges, farmland.
NOTES: Extending range northward. Stragglers are seen well to the north in the spring and fall.

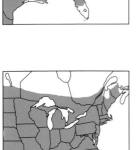

INDIGO BUNTING *Passerina cyanea* **5½"**
VOICE: A lively jumble of high notes: *set sweet sweet, chew chew, tree tree*.
HABITAT: Brushy pastures, power lines, scrub wood edges.
NOTES: Can look black if lit from behind. Rare north to Nova Scotia. Has hybridized with the Lazuli Bunting at western edge of range.

LAZULI BUNTING *Passerina amoena* **to 5½"**
VOICE: Similar to the Indigo's.
HABITAT: Open brush, streamside shrubs.
NOTES: Breeds to western edge of territory covered by this book. Its range overlaps with that of the Indigo, and it hybridizes regularly.

PAINTED BUNTING *Passerina ciris* **5⅕"**
VOICE: A pleasant warble. Call note is a *chip*.
HABITAT: Wood and dune edges, roadsides, suburban areas, and in towns, gardens. Visits feeders frequently.
NOTES: A visually arresting songbird. Casual to New England.

BLUE FINCHES, etc.

♀

♂

BLUE GROSBEAK

♂ molting

♂ summer

♀

INDIGO BUNTING

♀

♂

LAZULI BUNTING

♀

♂

PAINTED BUNTING

253

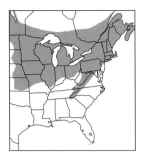

ROSE-BREASTED GROSBEAK
Pheucticus ludovicianus **to 8½"**
VOICE: A smooth rising and falling of mellow notes. Call note is a metallic *eek*.
HABITAT: Woodlands, orchards, parks.
NOTES: Rare at winter feeders. May hybridize with the Black-headed.

BLACK-HEADED GROSBEAK
Pheucticus melanocephalus **to 7⁷⁄₁₀"**
VOICE: Voice is like that of the Rose-breasted.
HABITAT: Strays to the East from the West. Thickets.
NOTES: Appears at feeders. Female has a yellow or buff chest with little streaking compared to the female Rose-breasted.

GREEN-TAILED TOWHEE
Pipilo chlorurus **6½"**
Another stray to the East from the West. Usually recorded at winter feeders. Has a white chin and a rich, rusty cap.

EASTERN TOWHEE
Pipilo erythrophthalmus **to 8½"**
VOICE: A distinct *drink-your-teeeee*, with the last syllable higher. Also a loud *chewink* with a rising inflection.
HABITAT: Dry open woods, bushy edges of undergrowth, pine barrens.
NOTES: White-eyed race in South. Uses both feet to double-scratch in leaves.

SPOTTED TOWHEE
Pipilo maculata **to 8½"**
VOICE: Call a slurred, whiny, scratchy *twee*.
HABITAT: Brushy areas.
NOTES: Occurs east to plains states.

GROSBEAKS,
TOWHEES

♂

ROSE-
BREASTED
GROSBEAK

♀

♀

♂

BLACK-HEADED
GROSBEAK

GREEN-TAILED
TOWHEE

♀

juvenile

white-eyed
race

♂

♂

EASTERN
TOWHEE

SPOTTED
TOWHEE

255

WHITE-THROATED SPARROW
Zonotrichia albicollis **to 7"**

VOICE: A clear, pensive, whistled *oh-sweet-sweet-sweet* or a double-parted *oh-sam-pea-body pea-body*. Call note is a hard *chink*.

HABITAT: Thickets, brush, bogs, marshes, wood edges. Frequents winter feeders.

NOTES: There are two forms: a white-striped head and a tan-striped head.

WHITE-CROWNED SPARROW
Zonotrichia leucophrys **to 7½"**

VOICE: A *sweet-sweet-sweet-teedle-de-de deet* that rises at the end.

HABITAT: Breeds in boreal scrub in East; brush edges, tangles, roadsides.

NOTES: Rarely travels to the southeastern coast.

HARRIS'S SPARROW
Zonotrichia querula **to 7½"**

VOICE: A quavering song of opening clear whistles on the same pitch, then lower-pitched notes. Alarm note is a *wink* (G. M. Sutton).

HABITAT: Breeds in stunted boreal forests and bogs. Brush, thickets, hedgerows, open woods in winter.

NOTES: This central plains species is rare to the East Coast, where it often visits feeders.

GOLDEN-CROWNED SPARROW
Zonotrichia atricapilla **to 7"**

This species is a casual to accidental visitor to the East. Most reports are of birds seen at feeding stations.

SPARROWS

WHITE-THROATED
SPARROW
tan-striped form

summer
similar in winter

white-striped form

summer
duller in winter

WHITE-CROWNED
SPARROW

immature

adult

immature

GOLDEN-
CROWNED
SPARROW

adult

adult
summer

immature
winter

HARRIS'S
SPARROW

CHIPPING SPARROW *Spizella passerina* 5"
VOICE: A staccato chipping on 1 pitch.
HABITAT: Towns, orchards, wood edges, gardens.
NOTES: Prefers to nest in evergreens.

FIELD SPARROW *Spizella pusilla* 5"
VOICE: A series of sweet notes that begins slowly, gains in speed, and ends in a trill.
HABITAT: Brushy fields, scrub.
NOTES: Often moves about in small groups after nesting season.

SWAMP SPARROW
Melospiza georgiana to 5⁷/₁₀"
VOICE: A strong, evenly spaced, rattled trill, at times on 2 pitches. Call note: *chip*.
HABITAT: Fresh marshes with tussocks, bushes, or cattails; sedgy swamps. Weedy fields in migration.
NOTES: Distinguish from Song Sparrow by sparsely streaked breast contrasting with whitish throat; gray eyebrow.

AMERICAN TREE SPARROW
Spizella arborea 6"
VOICE: A sweet jumble of notes—often compared to jingling of keys. Call note: *teelwit*.
HABITAT: Nests in scrub at Arctic treeline. Seen in scrub, in thickets, cattails, and at feeders in winter.
NOTES: Often forms large winter groups.

RUFOUS-CROWNED SPARROW
Aimophila ruficeps to 6"
VOICE: Stuttering, gurgles. Call note: *dear, dear, deur*.
HABITAT: Dry, open, brushy and rocky slopes.
NOTES: Breeds east locally to e. Oklahoma and w. Arkansas.

RUSTY-CAPPED SPARROWS

sexes similar

winter

summer

CHIPPING SPARROW

juvenile

for comparison with winter Chipping Sparrow, above

FIELD SPARROW

juvenile

CLAY-COLORED SPARROW
(see p. 261)

immature

SWAMP SPARROW

RUFOUS-CROWNED SPARROW

AMERICAN TREE SPARROW

259

LARK SPARROW
Chondestes grammacus   **to 6½"**
VOICE: Clear notes and trills with pauses between them. Buzzy phrases.
HABITAT: Open country with bushes, pastures, farms, roadsides.
NOTES: Has withdrawn from some breeding areas on eastern edge of range. A rare fall transient to the East Coast.

CLAY-COLORED SPARROW
Spizella pallida   **5"**
VOICE: Unbirdlike: 3 or 4 low, flat buzzes: *bzzz, bzzz, bzzz.*
HABITAT: Scrub, bushy prairies, jack pines.
NOTES: Pale lores. White mustache stripe. Extending its range eastward. A regular fall transient on the East Coast.

GRASSHOPPER SPARROW
Ammodramus savannarum   **to 5"**
VOICE: 2 faint introductory notes followed by high, thin buzz: *tsick, tsick-tiszeeeeeeeee.*
HABITAT: Grasslands, hay fields, prairies.
NOTES: Declining in many areas because of loss of grassland nesting habitat.

BACHMAN'S SPARROW
Aimophila aestivalis   **5⁷/₁₀"**
VOICE: A clear, liquid whistle followed by a loose trill or warble. (Vaguely suggests Hermit Thrush's song).
HABITAT: Open pine or oak woods with dense understory. Brushy pastures.
NOTES: Disappearing from northern parts of range. Populations are feeling pressure of habitat loss.

SPARROWS
(Mostly clear-breasted)

sexes similar

juvenile

adult

**LARK
SPARROW**

**CHIPPING
SPARROW**
winter

summer

winter

see summer adult
on p. 259

**CLAY-
COLORED
SPARROW**

**GRASSHOPPER
SPARROW**

see juvenile
on p. 265

adult

**BACHMAN'S
SPARROW**

**HENSLOW'S
SPARROW**

juvenile
see adult on p. 265

261

FOX SPARROW *Passerella iliaca* to 7½"
VOICE: Musical. A varied arrangement of short clear notes and whistles.
HABITAT: Wooded undergrowth, brush.
NOTES: A large sparrow that is a winter visitor to brushy areas and feeding stations.

SONG SPARROW *Melospiza melodia* to 6½"
VOICE: A series of rollicking notes. Clear notes to start: *Sweet-sweet-sweet*, etc.
HABITAT: Thickets, brush, roadsides, parks, gardens.
NOTES: Perhaps our most common sparrow. Has a long tail and a dark chest spot.

VESPER SPARROW *Pooecetes gramineus* 6"
VOICE: Throatier than Song Sparrow. Begins with 2 clear minor notes and then 2 higher ones.
HABITAT: Meadows, fields, prairies, roadsides.
NOTES: Disappearing from many areas of the Northeast because of loss of grassland habitat.

LINCOLN'S SPARROW
Melospiza lincolnii 5½"
VOICE: Sweet and gurgling, suggestive of the House Wren's voice. A low passage that rises, then drops.
HABITAT: Breeding grounds are willow and alder thickets, muskeg. Visits thickets and weedy sites in winter.
NOTES: This bird is secretive, so it is difficult to see it clearly during migration or in winter thickets.

FOX
SPARROW

SONG
SPARROW

VESPER
SPARROW

LINCOLN'S
SPARROW

SWAMP
SPARROW
(see p. 258)

immature

Purple
Finch
♀

House
Finch
♀

females for comparison with sparrows
(males shown on p. 249)

SAVANNAH SPARROW

Passerculus sandwichensis **to 5⁷⁄₁₀"**

VOICE: A lisping *tsit-tsit-tsit, teeeee-tsaaaay.*

HABITAT: Open fields, meadows, salt marsh edges, prairies, dunes, shoreline.

NOTES: Habitat loss is affecting some populations. A numerous sparrow in migration.

"IPSWICH" SAVANNAH SPARROW

Passerculus sandwichensis **6"**

VOICE: Similar to that of Savannah Sparrow.

HABITAT: Dunes, beach grass and marsh edge in winter.

NOTES: A large sandy gray race of the Savannah Sparrow with a restricted breeding range of Sable Island and a small area of coastal Nova Scotia. A winter visitor along the Atlantic Coast.

BAIRD'S SPARROW

Ammodramus bairdii **5"**

VOICE: 2 or 3 high musical *zips*, followed by a lower-pitched trill.

HABITAT: Local in northern high-grass prairies.

NOTES: An elusive and skulking species that is local in its distribu-tion.

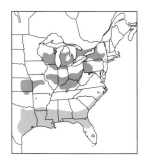

HENSLOW'S SPARROW

Ammodramus henslowii **to 5"**

VOICE: An insectlike *tis-lick.* Often sings all night.

HABITAT: Specific types of weedy and brushy fields.

NOTES: Colors of distinct olive tones. Mouselike and difficult to find. Henslow's Sparrow is very specific in its choice of breeding areas and has disappeared from most of the Northeast. Colonies are sporadic.

STREAK-BREASTED
SPARROWS

sexes similar

SAVANNAH
SPARROW
typical

SAVANNAH
SPARROW
dark northern form

"IPSWICH"
SAVANNAH
SPARROW

BAIRD'S
SPARROW

adult

HENSLOW'S
SPARROW

(juvenile on p. 261)

GRASSHOPPER
SPARROW

(adult on p. 261)

juvenile

SALTMARSH SHARP-TAILED SPARROW
Ammodramus caudacutus   **to 5½"**
VOICE: A gasping buzz: *tuptup-sheeeeee*.
HABITAT: Coastal marshes.
NOTES: Recent research indicates that Nelson's race is a full species.

NELSON'S SHARP-TAILED SPARROW
Ammodramus nelsoni   **to 5½"**
VOICE: Loud, hissing buzz: *seep-sssssssss*.
HABITAT: Tall grasses of freshwater and some saltwater marsh.
NOTES: Recent split from Saltmarsh Sharp-tailed Sparrow. Occurs in coastal marshes during migration and winter.

Lᴇ CONTE'S SPARROW
Ammodramus leconteii   **to 5⅕"**
VOICE: 2 thin, grasshopper-like hisses.
HABITAT: Tall grass, weedy fields, marsh edges.
NOTES: LeConte's has a white stripe over the crown and an orangey-yellow eyeline. A skulker, it is difficult to see. Occasionally wanders to the East Coast.

SEASIDE SPARROW
Ammodramus maritimus   **to 6"**
VOICE: A short intro of 2 notes, then a buzz: *cut cut-zhe-eeeeeee*.
HABITAT: Salt marshes.
NOTES: Large and dark. Best seen in morning when singing on salt marsh grasses.

"DUSKY" SEASIDE SPARROW and "CAPE SABLE" SEASIDE SPARROW
Both subspecies of the Seaside Sparrow. The Dusky was confined to the Titusville area of Florida, and habitat loss caused its extinction in 1987. The Cape Sable form holds on in the Cape Sable area and local sites in the Everglades. The population density is low, and it has been placed on the endangered list because of habitat loss.

MARSH SPARROWS

typical form

maritime ("Acadian") race

sexes similar

SALTMARSH SHARP-TAILED SPARROW

NELSON'S SHARP-TAILED SPARROW

LE CONTE'S SPARROW

SEASIDE SPARROW

juvenile

"CAPE SABLE" SEASIDE SPARROW

"DUSKY" SEASIDE SPARROW

LIFE LIST

Keep a Life List. Check the birds you have seen. This list covers the eastern half of the continent west to the 100th meridian on the Great Plains.

For a convenient and complete continental list, the A.B.A. Check-list, prepared by the Check-list Committee of the American Birding Association (Box 6599, Colorado Springs, Colorado 80934 or www.american-birding.org), is recommended. It lists every species recorded north of the Mexican border, including accidentals.

The order of this list differs from the order of the plates in the book; the list follows the taxonomic order set out in the 1998 ABA Check-list. In this Life List, birds are grouped first under orders (identified by the Latin ending *-formes*), followed by families (*-dae* ending), sometimes subfamilies (*-nae* ending), and then species. Scientific names of genera and species are not given below but will be found in the species accounts throughout the book.

ORDER GAVIIFORMES
LOONS: Gaviidae
____Red-throated Loon
____Pacific Loon
____Common Loon
____Yellow-billed Loon

ORDER PODICIPEDIFORMES
GREBES: Podicipedidae
____Pied-billed Grebe
____Horned Grebe
____Red-necked Grebe
____Eared Grebe
____Western Grebe
____Clark's Grebe

ORDER PROCELLARIIFORMES
SHEARWATERS, PETRELS: Procellariidae
____Northern Fulmar
____Bermuda Petrel
____Black-capped Petrel
____Cory's Shearwater
____Greater Shearwater
____Sooty Shearwater
____Manx Shearwater

____Audubon's Shearwater
STORM-PETRELS: Hydrobatidae
____Wilson's Storm-Petrel
____Leach's Storm-Petrel

ORDER PELECANIFORMES
TROPICBIRDS: Phaethontidae
____White-tailed Tropicbird
BOOBIES, GANNETS: Sulidae
____Masked Booby
____Brown Booby
____Northern Gannet
PELICANS: Pelecanidae
____American White Pelican
____Brown Pelican
CORMORANTS: Phalacrocoracidae
____Neotropic Cormorant
____Double-crested Cormorant
____Great Cormorant
DARTERS: Anhingidae
____Anhinga
FRIGATEBIRDS: Fregatidae
____Magnificent Frigatebird

ORDER CICONIIFORMES
HERONS, BITTERNS: Ardeidae
____American Bittern

____Least Bittern
____Great Blue Heron
    ____"Wurdemann's" Heron
    ____"Great White" Heron
____Great Egret
____Snowy Egret
____Little Blue Heron
____Tricolored Heron
____Reddish Egret
____Cattle Egret
____Green Heron
____Black-crowned Night-Heron
____Yellow-crowned Night-Heron

IBISES, SPOONBILLS: Threskiornithidae
____White Ibis
____Scarlet Ibis
____Glossy Ibis
____White-faced Ibis
____Roseate Spoonbill

STORKS: Ciconiidae
____Wood Stork

AMERICAN VULTURES: Cathartidae
____Black Vulture
____Turkey Vulture

FLAMINGOS: Phoenicopteridae
____Greater Flamingo

ORDER ANSERIFORMES
WATERFOWL: Anatidae
Whistling-Ducks: Dendrocygninae
____Fulvous Whistling-Duck

Geese and Swans: Anserinae
____Greater White-fronted Goose
____Snow Goose
    ____Blue Goose
____Ross's Goose
____Canada Goose
____Brant
____Barnacle Goose
____Mute Swan
____Tundra Swan

True Ducks: Anatinae
____Muscovy Duck
____Wood Duck
____Gadwall
____Eurasian Wigeon
____American Wigeon
____American Black Duck
____Mallard

____Mottled Duck
____Blue-winged Teal
____Cinnamon Teal
____Northern Shoveler
____Northern Pintail
____Green-winged Teal
    ____"Eurasian" Teal
____Canvasback
____Redhead
____Ring-necked Duck
____Tufted Duck
____Greater Scaup
____Lesser Scaup
____King Eider
____Common Eider
____Harlequin Duck
____Surf Scoter
____White-winged Scoter
____Black Scoter
____Oldsquaw
____Bufflehead
____Common Goldeneye
____Barrow's Goldeneye
____Hooded Merganser
____Common Merganser
____Red-breasted Merganser
____Ruddy Duck

ORDER FALCONIFORMES
HAWKS, EAGLES, etc.: Accipitridae
Ospreys: Pandionidae
____Osprey

Kites, Eagles, Hawks: Accipitrinae
____Swallow-tailed Kite
____White-tailed Kite
____Snail Kite
____Mississippi Kite
____Bald Eagle
____Northern Harrier
____Sharp-shinned Hawk
____Cooper's Hawk
____Northern Goshawk
____Red-shouldered Hawk
____Broad-winged Hawk
____Short-tailed Hawk
____Swainson's Hawk
____Red-tailed Hawk
    ____"Harlan's" Red-tailed Hawk
    ____"Krider's" Red-tailed Hawk
____Ferruginous Hawk

269

____Rough-legged Hawk
____Golden Eagle

CARACARAS, FALCONS: Falconidae
Caracaras: Caracarinae
____Crested Caracara
Falcons: Falconinae
____American Kestrel
____Merlin
____Gyrfalcon
____Peregrine Falcon
____Prairie Falcon

ORDER GALLIFORMES
PARTRIDGES, GROUSE, TURKEYS, QUAILS, : Phasianidae
Partridges, Pheasants: Phasianinae
____Gray Partridge
____Ring-necked Pheasant
Grouse: Tetraoninae
____Ruffed Grouse
____Spruce Grouse
____Willow Ptarmigan
____Rock Ptarmigan
____Sharp-tailed Grouse
____Greater Prairie-Chicken
____Lesser Prairie-Chicken
Turkeys: Meleagridinae
____Wild Turkey

ORDER ODONTOPHORIDAE
____Northern Bobwhite
____Scaled Quail

ORDER GRUIFORMES
RAILS, GALLINULES, COOTS: Rallidae
____Yellow Rail
____Black Rail
____Clapper Rail
____King Rail
____Virginia Rail
____Sora
____Purple Gallinule
____Common Moorhen
____American Coot
LIMPKINS: Aramidae
____Limpkin
CRANES: Gruidae
____Sandhill Crane
____Whooping Crane

ORDER CHARADRIIFORMES
Plovers: Charadriinae
____Black-bellied Plover
____American Golden-Plover
____Snowy Plover
____Wilson's Plover
____Ringed Plover
____Semipalmated Plover
____Piping Plover
____Killdeer
OYSTERCATCHERS: Haematopodidae
____American Oystercatcher
STILTS, AVOCETS: Recurvirostridae
____Black-necked Stilt
____American Avocet
SANDPIPERS, PHALAROPES, etc. Scolopacidae
Sandpipers and Allies: Scolopacinae
____Greater Yellowlegs
____Lesser Yellowlegs
____Solitary Sandpiper
____Willet
____Spotted Sandpiper
____Upland Sandpiper
____Whimbrel
____Long-billed Curlew
____Hudsonian Godwit
____Marbled Godwit
____Ruddy Turnstone
____Red Knot
____Sanderling
____Semipalmated Sandpiper
____Western Sandpiper
____Least Sandpiper
____White-rumped Sandpiper
____Baird's Sandpiper
____Pectoral Sandpiper
____Purple Sandpiper
____Dunlin
____Curlew Sandpiper
____Stilt Sandpiper
____Buff-breasted Sandpiper
____Ruff
____Short-billed Dowitcher
____Long-billed Dowitcher
____Common Snipe
____American Woodcock
Phalaropes: Phalaropodinae
____Wilson's Phalarope

____Red-necked Phalarope
____Red Phalarope

SKUAS, GULLS, TERNS, SKIMMERS: Laridae

Skuas, Jaegers: Stercorariinae
____Great Skua
____South Polar Skua
____Pomarine Jaeger
____Parasitic Jaeger
____Long-tailed Jaeger

Gulls: Larinae
____Laughing Gull
____Franklin's Gull
____Little Gull
____Black-headed Gull
____Bonaparte's Gull
____Ring-billed Gull
____California Gull
____Herring Gull
____Thayer's Gull
____"Kumlien's" Iceland Gull
____Lesser Black-backed Gull
____Glaucous Gull
____Great Black-backed Gull
____Sabine's Gull
____Black-legged Kittiwake
____Ross's Gull
____Ivory Gull

Terns: Sterninae
____Gull-billed Tern
____Caspian Tern
____Royal Tern
____Sandwich Tern
____Roseate Tern
____Common Tern
____Arctic Tern
____Forster's Tern
____Least Tern
____Bridled Tern
____Sooty Tern
____Black Tern
____Brown Noddy
____Black Noddy

Skimmers: Rynchopinae
____Black Skimmer

AUKS, MURRES, PUFFINS: Alcidae
____Dovekie
____Common Murre
____Thick-billed Murre
____Razorbill
____Black Guillemot
____Atlantic Puffin

ORDER COLUMBIFORMES
PIGEONS, DOVES: Columbidae
____Rock Dove
____White-crowned Pigeon
____Ringed Turtle-Dove
____White-winged Dove
____Mourning Dove
____Inca Dove
____Common Ground-Dove

ORDER PSITTACIFORMES
PARAKEETS, PARROTS: Psittacidae
Australian Parakeets, Rosellas: Platycercinae
____Budgerigar
Macaws, Parrots: Arinae
____Monk Parakeet
____White-winged Parakeet

ORDER CUCULIFORMES
CUCKOOS, ROADRUNNERS, AND ALLIES: Cuculidae
New World Cuckoos: Coccyzinae
____Black-billed Cuckoo
____Yellow-billed Cuckoo
____Mangrove Cuckoo
Roadrunners: Neomorphinae
____Greater Roadrunner
Anis: Crotophaginae
____Smooth-billed Ani
____Groove-billed Ani

ORDER STRIGIFORMES
BARN OWLS: Tytonidae
____Barn Owl
TYPICAL OWLS: Strigidae
____Eastern Screech-Owl
____Great Horned Owl
____Snowy Owl
____Northern Hawk Owl
____Burrowing Owl
____Barred Owl
____Great Gray Owl
____Long-eared Owl
____Short-eared Owl
____Boreal Owl
____Northern Saw-whet Owl

ORDER CAPRIMULGIFORMES
GOATSUCKERS: Caprimulgidae
Nighthawks: Chordeilinae
____Common Nighthawk
Nightjars: Caprimulginae
____Common Poorwill
____Chuck-will's-widow
____Whip-poor-will

ORDER APODIFORMES
SWIFTS: Apodidae
Chaeturine Swifts: Chaeturinae
____Chimney Swift
____Vaux's Swift
HUMMINGBIRDS: Trochilidae
Typical Hummingbirds: Trochilinae
____Ruby-throated Hummingbird
____Rufous Hummingbird

ORDER CORACIIFORMES
KINGFISHERS: Alcedinidae
Typical Kingfishers: Cerylinae
____Belted Kingfisher

ORDER PICIFORMES
WOODPECKERS: Picidae
____Red-headed Woodpecker
____Red-bellied Woodpecker
____Yellow-bellied Sapsucker
____Downy Woodpecker
____Hairy Woodpecker
____Red-cockaded Woodpecker
____Three-toed Woodpecker
____Black-backed Woodpecker
____Northern Flicker (Yellow-shafted)
____"Red-shafted Flicker"
____Pileated Woodpecker

ORDER PASSERIFORMES
TYRANT FLYCATCHERS: Tyrannidae
Fluvicoline Flycatchers: Fluvicolinae
____Olive-sided Flycatcher
____Eastern Wood-Pewee
____Yellow-bellied Flycatcher
____Acadian Flycatcher
____Alder Flycatcher
____Willow Flycatcher
____Least Flycatcher
____Eastern Phoebe
____Say's Phoebe
____Vermilion Flycatcher

Tyrannine Flycatchers: Tyranninae
____Ash-throated Flycatcher
____Great Crested Flycatcher
____Western Kingbird
____Eastern Kingbird
____Gray Kingbird
____Scissor-tailed Flycatcher
SHRIKES: Laniidae
____Loggerhead Shrike
____Northern Shrike
VIREOS: Vireonidae
____White-eyed Vireo
____Bell's Vireo
____Black-capped Vireo
____Yellow-throated Vireo
____Blue-headed Vireo
____Warbling Vireo
____Black-whiskered Vireo
____Philadelphia Vireo
____Red-eyed Vireo
CROWS, JAYS: Corvidae
____Gray Jay
____Blue Jay
____Florida Scrub-Jay
____Black-billed Magpie
____American Crow
____Common Raven
____Fish Crow
____Chihuahuan Raven
LARKS: Alaudidae
____Horned Lark
SWALLOWS: Hirundinidae
Typical Swallows: Hirundininae
____Purple Martin
____Tree Swallow
____Northern Rough-winged Swallow
____Bank Swallow
____Cliff Swallow
____Barn Swallow
CHICKADEES, TITMICE: Paridae
____Carolina Chickadee
____Black-capped Chickadee
____Boreal Chickadee
____Tufted Titmouse
NUTHATCHES: Sittidae
Nuthatches: Sittinae
____Red-breasted Nuthatch
____White-breasted Nuthatch
____Brown-headed Nuthatch

CREEPERS: Certhiidae
Northern Creepers: Certhiinae
____Brown Creeper

WRENS: Troglodytidae
____Rock Wren
____Carolina Wren
____Bewick's Wren
____House Wren
____Winter Wren
____Sedge Wren
____Marsh Wren

BULBULS: Pycnonotidae
____Red-whiskered Bulbul

KINGLETS: Regulidae
____Golden-crowned Kinglet
____Ruby-crowned Kinglet

GNATCATCHERS: Sylviidae
Gnatcatchers: Polioptilinae
____Blue-gray Gnatcatcher

THRUSHES: Turdidae
____Northern Wheatear
____Eastern Bluebird
____Mountain Bluebird
____Townsend's Solitaire
____Veery
____Gray-cheeked Thrush
____Bicknell's Thrush
____Swainson's Thrush
____Hermit Thrush
____Wood Thrush
____American Robin
____Varied Thrush

MOCKINGBIRDS, THRASHERS: Mimidae
____Gray Catbird
____Northern Mockingbird
____Brown Thrasher

STARLINGS: Sturnidae
____European Starling

PIPITS: Motacillidae
____American Pipit
____Sprague's Pipit

WAXWINGS: Bombycillidae
____Bohemian Waxwing
____Cedar Waxwing

WOOD WARBLERS: Parulidae
____Blue-winged Warbler
    ____"Brewster's" Warbler (hybrid)
    ____"Lawrence's" Warbler (hybrid)

____Golden-winged Warbler
____Tennessee Warbler
____Orange-crowned Warbler
____Nashville Warbler
____Northern Parula
    ____"Sutton's" Warbler (hybrid)
____Yellow Warbler
____Chestnut-sided Warbler
____Magnolia Warbler
____Cape May Warbler
____Black-throated Blue Warbler
____Yellow-rumped Warbler
    ____"Myrtle" Warbler
    ____"Audubon's" Warbler
____Black-throated Gray Warbler
____Black-throated Green Warbler
____Blackburnian Warbler
____Yellow-throated Warbler
____Pine Warbler
____Kirtland's Warbler
____Prairie Warbler
____Palm Warbler
____Bay-breasted Warbler
____Blackpoll Warbler
____Cerulean Warbler
____Black-and-white Warbler
____American Redstart
____Prothonotary Warbler
____Worm-eating Warbler
____Swainson's Warbler
____Ovenbird
____Northern Waterthrush
____Louisiana Waterthrush
____Kentucky Warbler
____Connecticut Warbler
____Mourning Warbler
____Common Yellowthroat
____Hooded Warbler
____Wilson's Warbler
____Canada Warbler
____Yellow-breasted Chat

TANAGERS: Thraupidae
____Summer Tanager
____Scarlet Tanager
____Western Tanager
____Blue-Gray Tanager

EMBERIZIDS: Emberizidae
____Green-tailed Towhee
____Spotted Towhee
____Eastern Towhee
____Bachman's Sparrow

____Rufous-crowned Sparrow
____American Tree Sparrow
____Chipping Sparrow
____Clay-colored Sparrow
____Field Sparrow
____Vesper Sparrow
____Lark Sparrow
____Lark Bunting
____Savannah Sparrow
    ____"Ipswich" Sparrow
____Grasshopper Sparrow
____Baird's Sparrow
____Henslow's Sparrow
____Le Conte's Sparrow
____Nelson's Sharp-tailed Sparrow
____Saltmarsh Sharp-tailed Sparrow
____Seaside Sparrow
    ____"Cape Sable" Seaside Sparrow
____Fox Sparrow
____Song Sparrow
____Lincoln's Sparrow
____Swamp Sparrow
____White-throated Sparrow
____Harris's Sparrow
____White-crowned Sparrow
____Golden-crowned Sparrow
____Dark-eyed Junco
    ____"Slate-colored" Junco
    ____"Oregon" Junco
    ____"White-winged" Junco
____McCown's Longspur
____Lapland Longspur
____Smith's Longspur
____Chestnut-collared Longspur
____Snow Bunting

CARDINALS AND ALLIES: Cardinalidae
____Northern Cardinal
____Rose-breasted Grosbeak
____Black-headed Grosbeak

____Blue Grosbeak
____Lazuli Bunting
____Indigo Bunting
____Painted Bunting
____Dickcissel

BLACKBIRDS: Icteridae
____Bobolink
____Red-winged Blackbird
____Eastern Meadowlark
____Western Meadowlark
____Yellow-headed Blackbird
____Rusty Blackbird
____Brewer's Blackbird
____Common Grackle
____Boat-tailed Grackle
____Great-tailed Grackle
____Brown-headed Cowbird
____Orchard Oriole
____Spot-breasted Oriole
____Baltimore Oriole
____Bullock's Oriole

FINCHES AND ALLIES: Fringillidae
Cardueline Finches: Carduelinae
____Pine Grosbeak
____Purple Finch
____House Finch
____Red Crossbill
____White-winged Crossbill
____Common Redpoll
____Hoary Redpoll
____Pine Siskin
____American Goldfinch
____European Goldfinch
____Evening Grosbeak

OLD WORLD SPARROWS: Passeridae
____House Sparrow
____Eurasian Tree Sparrow

INDEX

All birds illustrated and described in this book are indexed. The page number(s), with rare exception, refer(s) to the text; it is understood that the illustration is on the right-hand facing page. Scientific names (in *italics*) are keyed to the pages on which the text appears, not the illustrations.

277

285

287

BIRDS

ADVANCED BIRDING (39) North America 97500-X

BIRDS OF BRITAIN AND EUROPE (8) 66922-7

BIRDS OF TEXAS (13) Texas and adjacent states 92138-4

BIRDS OF THE WEST INDIES (18) 0-618-00210-3

EASTERN BIRDS (1) Eastern and central North America 91176-1

EASTERN BIRDS' NESTS (21) U.S. east of Mississippi River 93609-8

HAWKS (35) North America 93615-2

WESTERN BIRDS (2) North America west of 100th meridian and north of Mexico 91173-7

WESTERN BIRDS' NESTS (25) U.S. west of Mississippi River 47863-4

MEXICAN BIRDS (20) Mexico, Guatemala, Belize, El Salvador 97514-X

WARBLERS (49) North America 78321-6

FISH

PACIFIC COAST FISHES (28) Gulf of Alaska to Baja California 0-618-00212-X

ATLANTIC COAST FISHES (32) North American Atlantic coast 97515-8

FRESHWATER FISHES (42) North America north of Mexico 91091-9

INSECTS

INSECTS (19) North America north of Mexico 91170-2

BEETLES (29) North America 91089-7

EASTERN BUTTERFLIES (4) Eastern and central North America 90453-6

WESTERN BUTTERFLIES (33) U.S. and Canada west of 100th meridian, part of northern Mexico 79151-0

MAMMALS

MAMMALS (5) North America north of Mexico 91098-6

ANIMAL TRACKS (9) North America 91094-3

ECOLOGY

EASTERN FORESTS (37) Eastern North America 9289-5

CALIFORNIA AND PACIFIC NORTHWEST FORESTS (50) 92896-6

ROCKY MOUNTAIN AND SOUTHWEST FORESTS (51) 92897-4

VENOMOUS ANIMALS AND POISONOUS PLANTS (46) North America north of Mexico 93608-X

PLANTS

EDIBLE WILD PLANTS (23) Eastern and central North America 92622-X

EASTERN TREES (11) North America east of 100th meridian 90455-2

FERNS (10) Northeastern and central North America, British Isles and Western Europe 97512-3

MEDICINAL PLANTS (40) Eastern and central North America 92066-3

MUSHROOMS (34) North America 91090-0

PACIFIC STATES WILDFLOWERS (22) Washington, Oregon, California, and adjacent areas 91095-1

ROCKY MOUNTAIN WILDFLOWERS (14) Northern Arizona and New Mexico to British Columbia 93613-6

TREES AND SHRUBS (11A) Northeastern and north-central U.S. and southeastern and south-central Canada 35370-X

WESTERN TREES (44) Western U.S. and Canada 90454-4

WILDFLOWERS OF NORTHEASTERN AND NORTH-CENTRAL NORTH AMERICA (17) 91172-9

SOUTHWEST AND TEXAS WILDFLOWERS (31) 93612-8

EARTH AND SKY

GEOLOGY (48) Eastern North America 66326-1

ROCKS AND MINERALS (7) North America 91096-X

STARS AND PLANETS (15) 93431-1

ATMOSPHERE (26) 97631-6

REPTILES AND AMPHIBIANS

EASTERN REPTILES AND AMPHIBIANS (12) Eastern and central North America 90452-8

WESTERN REPTILES AND AMPHIBIANS (16) Western North America, including Baja California 93611-X

SEASHORE

SHELLS OF THE ATLANTIC (3) Atlantic and Gulf coasts and the West Indies 69779-4

PACIFIC COAST SHELLS (6) North American Pacific coast, including Hawaii and the Gulf of California 18322-7

ATLANTIC SEASHORE (24) Bay of Fundy to Cape Hatteras 0-618-00209-X

CORAL REEFS (27) Caribbean and Florida 0-618-00211-1

SOUTHEAST AND CARIBBEAN SEASHORES (36) Cape Hatteras to the Gulf Coast, Florida, and the Caribbean 97516-6

PETERSON
NATURAL HISTORY COMPANIONS
LIVES OF NORTH AMERICAN BIRDS 77017-3

PETERSON FIRST GUIDES®

ASTRONOMY 935423
BIRDS 90666-0
BUTTERFLIES AND MOTHS 90665-2
CATERPILLARS 91184-2
CLOUDS AND WEATHER 90993-6
DINOSAURS 97196-9
FISHES 91179-6
INSECTS 90664-4
MAMMALS 91181-8
REPTILES AND AMPHIBIANS 97195-0
ROCKS AND MINERALS 93543-1
SEASHORES 91180-X
SHELLS 91182-6
SOLAR SYSTEM 97194-2
TREES 91183-4
URBAN WILDLIFE 93544-X
WILDFLOWERS 90667-9
FORESTS 97197-7

AUDIO AND VIDEO

EASTERN BIRDING BY EAR
cassettes 97523-9
CD 97524-7
WESTERN BIRDING BY EAR
cassettes 97526-3
CD 97525-5
EASTERN BIRD SONGS, Revised
cassettes 53150-0
CD 97522-0
WESTERN BIRD SONGS, Revised
cassettes 51746-X
CD 975190
EASTERN MORE BIRDING BY EAR
cassettes 97529-8
CD 97530-1
BACKYARD BIRD SONG
cassettes 97527-1
CD 97528-X

**PETERSON'S MULTIMEDIA GUIDES:
NORTH AMERICAN BIRDS**
(CD-ROM for Windows) 73056-2

PETERSON FIELD GUIDE
COLORING BOOKS

BIRDS 32521-8
BUTTERFLIES 34675-4
DESERTS 67086-1
DINOSAURS 49323-4
ENDANGERED WILDLIFE 57324-6
FISHES 44095-5
FORESTS 34676-2
INSECTS 67088-8
MAMMALS 44091-2
REPTILES 37704-8
SEASHORES 49324-2
SHELLS 37703-X
TROPICAL FORESTS 57321-1
WILDFLOWERS 32522-6

PETERSON FLASHGUIDES™

ATLANTIC COASTAL BIRDS 79286-X
PACIFIC COASTAL BIRDS 79287-8
EASTERN TRAILSIDE BIRDS 79288-6
WESTERN TRAILSIDE BIRDS 79289-4
HAWKS 79291-6
BACKYARD BIRDS 79290-8
TREES 82998-4
MUSHROOMS 82999-2
ANIMAL TRACKS 82997-6
BUTTERFLIES 82996-8
ROADSIDE WILDFLOWERS 82995-X
BIRDS OF THE MIDWEST 86733-9
WATERFOWL 86734-7
FRESHWATER FISHES 86713-4

WORLD WIDE WEB: http://www.petersononline.com

PETERSON FIELD GUIDES can be purchased at your local bookstore or by calling our toll-free number, (800) 225-3362.

When referring to title by corresponding ISBN number, preface with 0-395 unless title is listed with 0-618.

SWIMMERS

AERIALISTS

LONG-LEGGED WADERS

SMALLER WADERS

FOWL-LIKE BIRDS

BIRDS OF PREY

NONPASSERINE LAND BIRDS

PASSERINE (PERCHING) BIRDS